Regulating Technological Innovation

Regulating Technological Innovation

A Multidisciplinary Approach

Edited by

Michiel A. Heldeweg
Professor of Public Governance Law, University of Twente, the Netherlands

and

Evisa Kica
Researcher in Governance and Innovation, University of Twente, the Netherlands

First published 2011 by
PALGRAVE MACMILLAN

Palgrave Macmillan in the UK is an imprint of Macmillan Publishers Limited, registered in England, company number 785998, of Houndmills, Basingstoke, Hampshire RG21 6XS.

Palgrave Macmillan in the US is a division of St Martin's Press LLC, 175 Fifth Avenue, New York, NY 10010.

Palgrave Macmillan is the global academic imprint of the above companies and has companies and representatives throughout the world.

Palgrave® and Macmillan® are registered trademarks in the United States, the United Kingdom, Europe and other countries.

ISBN 978–0–230–36363–2

This book is printed on paper suitable for recycling and made from fully managed and sustained forest sources. Logging, pulping and manufacturing processes are expected to conform to the environmental regulations of the country of origin.

A catalogue record for this book is available from the British Library.

A catalog record for this book is available from the Library of Congress.

10   9   8   7   6   5   4   3   2   1
20   19   18   17   16   15   14   13   12   11

Printed and bound in Great Britain by
CPI Antony Rowe, Chippenham and Eastbourne

Contents

List of Tables

Notes on Contributors

Antoni J. P. Brack is Professor of Business Legal Studies at the University of Twente, the Netherlands. He has authored a book on managerial law (Bedrijfsrecht, 5th edr., 2010) and published numerous articles including in the *European Business Law Journal*. He is founder and co-editor of the *Dutch Consumer Law Journal*. He is a non-executive director of a UK-owned Dutch chemical company and a part-time judge in the Court of Appeal at Arnhem, the Netherlands.

Lesley C. P. Broos is a researcher and lecturer in the combined field of IT/telecommunications and law at the University of Twente, the Netherlands. He is currently preparing a PhD thesis on enhancing technological innovation in telecommunication infrastructure services and has published on the issues of regulatory supervision and first and second mover advantage theory relating to technology advancement.

Nupur Chowdhury is a PhD student in law at the School of Management and Governance, University of Twente. She is researching the issue of legal certainty in the context of a multilevel product quality and safety regulation of medical products.

Shawn Donnelly is an assistant professor of European Economic Governance at the University of Twente, the Netherlands. He is the author of *Reshaping Economic and Monetary Union* (Manchester 2004) and *The Regimes of European Integration: Constructing Governance of the Single Market* (Oxford 2010).

Michiel A. Heldeweg is Professor of Public Governance Law at the University of Twente and an honorary judge (within the Dutch District Court system; administrative bench). His research is on 'smart rules and regimes', fostering technological innovation while protecting against related risks, as set out in his 2009 inaugural lecture. His focus lies on regulation in the field of Environmental and Energy Law ('Sustainable Energy Law'). He is especially interested in matters of legal design and good legal governance in the new or post-regulatory state.

Martin W. Holterman is a post-doctoral researcher at the European University Institute. Previously, he was a PhD student at the University of Twente, the Netherlands. His research includes a theoretical analysis of the problems of semi-public institutions applying neo-institutional economics, and when such a set up is optimal.

Evisa Kica is a PhD student at the University of Twente. Her research focuses on the legitimacy of the multilevel standardisation process as it relates to the field of nanotechnology. In addition, she has been involved in EU contract research projects related to patent quality and innovation issues, and has investigated the role of patents in emerging technologies.

Luisa Marin is an assistant professor of European Law at the University of Twente, the Netherlands. Her research focuses on the constitutional analysis of the policies and instruments of the EU's Area of Freedom, Security and Justice, with particular attention to European criminal law cooperation. She has written on the European Arrest Warrant and its implementation in domestic legal Orders and co-edited the book *Still not resolved? Constitutional Challenges to the European Arrest Warrant*, Nijmegen, Wolf Legal Publishers, 2009, together with Elspeth Guild. Her recent research covers migration and border controls and Internet law.

Victor Rodriguez is an assistant professor in the Department of Legal and Economic Governance Studies at the University of Twente. He lectures on European Economic Governance and Macroeconomics. He has extensively researched the effects of biological material transfer agreements and has been involved in European projects on patents, corporate R&D and innovation. Previously, he was a tenured researcher in the TNO Innovation Policy Group and a visiting scholar at the University of Ottawa and the Conference Board of Canada.

Maurits P. T. Sanders studied public administration at the University of Twente in the Netherlands. He is a PhD student at the University of Twente while working as a lecturer in Public Policy Making at Saxion Universities of Applied Sciences in the Netherlands. His PhD research is on complex decision making in public-private partnerships.

Ramses A. Wessel is Professor of the Law of the European Union and other International Organisations at the University of Twente, the Netherlands. His research focuses on the relationship between international, EU and national laws. He is the editor of the *International Organizations Law Review* and in 2008 co-edited the book *Multilevel Regulation and the EU: The Interplay between Global, European and National Normative Processes* (The Hague/Boston: Martinus Nijhoff Publishers).

Introduction

Michiel A. Heldeweg and Evisa Kica

This book consists of a collection of essays concerning the relationship between regulation and technological innovation. The book represents an academic exchange of ideas on the realities and challenges of regulating technological innovation. It examines the regulatory issues surrounding the fostering technological innovation and its applications, and combines legal, economic and administrative science perspectives.

In particular, this book highlights the answers to important questions such as what type of regulatory framework would best fit the needs of technology and innovation developments; what competences or authorities should be given to the regulatory actors and other stakeholders to shape the future paths of technology innovation; what lessons can we distil from other regulatory fields; and how we can apply what we have learnt to further enhance the development of technology innovation?

I.1 Scope

This book starts from the European ambition (both at the European level and 'below') to become a knowledge-based economy and to secure technological innovation. Against this backdrop, the book aims to support and elucidate regulatory relevance as regards fostering innovation, while at the same time considering the regulatory need to strike a balance between fostering innovation and protecting against technological risks. To this end, a multidisciplinary perspective is applied as the role of regulation in the challenge of innovation is a multifaceted issue that cannot be properly understood from a single disciplinary angle.

Various documents and agendas, including the Lisbon agenda, and the specific multilevel governance characteristics of Europe, comprising both a supranational regulatory system, and (trans)national and regional diversities (in economic, cultural, social and other aspects and dynamics), set the stage for addressing the book's aims. The book is meant primarily to be relevant to the overall European context, although some contributions will address more 'local' governance particularities.

While regulating technological risks is a well-researched issue, this book aspires to understand better how regulation can effectively foster and secure technological innovation (while enhancing and protecting prerequisites and possible outcomes; and balancing technology development against risk regulation). As such, its attention is drawn to the field of regulation for fostering innovation, an area that is scientifically much less explored and understood. Beyond calls for deregulation and reducing the regulatory (and accompanying administrative) burden, regulation has a positive role to play in innovation (and growth), and is indeed indispensible in enhancing and securing both scientific exploration and, especially, its uptake and exploitation by the market.

In accentuating the 'fostering' of technological innovation, this book adopts three approaches. The first aims to provide a general analysis and appraisal of the relationship between regulation and innovation (mostly viewed in terms of comparative advantages, legal designs and informal regulation). The second addresses specific exemplary areas and connected issues in regulating technological innovation (new telecommunication infrastructures and/or related services, competition law and regulated industries, border management, energy innovation and Public Private Partnerships (PPP)). The third, and our last, approach focuses on emerging technologies and the regulation of innovation (addressing regulatory partnerships in nanotechnology, patent quality and the use of existing patents). Within these approaches, the reader will inevitably find a mix of concepts, values, strategies and practices for regulatory governance (both public and private) as potential ways to facilitate, enhance and secure technological innovation.

The underlying premise of the book is that technological innovation is a public value, either for reasons of fostering technological advancement (broadly or connected to specific societal needs and services) or for reasons of public risk control. Thus, the regulatory standpoint regarding innovation ultimately rests on the need to ensure societal opportunities, needs and risks related to a technological innovation are met. This does not mean, however, that only public or formal regulation (and public, second or third-party regulators) is relevant to the book. Many contributions explicitly show how the interplay between public and private actors, and public and private regulation, is a key characteristic of regulatory innovation governance. Further, the much sought after 'smart regulation' – presently a focal issue in EU regulatory governance – which features the idea of a combination of regulatory instruments, starts from the notion that there is no one-size-fits-all approach to combining regulation and innovation.

The science-based approach to the subject, as presented in this book, goes beyond showing that regulation matters in enhancing innovation. The book aims to highlight methods and techniques, and the accompanying pitfalls, that are relevant in defining and securing a proper balance between risks and opportunities, and between the public and private interests involved. The three approaches outlined above are intended to support this objective: they should demonstrate relevance and, moreover, present specific methodological insights into the specific issues of each approach (from general instrument choices and approaches, to addressing specific sectors or interests and, finally, how to deal with emerging technologies). Whether the results from the individual contributions can be combined to produce a 'smart design' for regulating innovation remains to be seen but, hopefully, this book will provide a step in that direction. In terms of scope, the contributions are geared to advancing this perspective, especially with regards to the European discourse and practice of regulating innovation.

I.2   Literature

Regulation of markets, particularly financial ones, and of public risks has received considerable attention in the literature throughout the previous century and into the current one.[1] Further, regulation for innovation has attracted the attention of various scholars following developments in emerging technologies that have associated risks and high levels of uncertainty.

In this book, the authors build on the definition of regulation formulated by Julia Black: 'regulation is the sustained and focused attempt to alter the behaviour of others according to standards or goals with the intention of producing a broadly identified outcome or outcomes, which may involve mechanisms of standard-setting, information-gathering and behaviour-modification'.[2]

Studies on the regulation and innovation relationship have enjoyed intensive debate among scholars. The main focus, however, has been on risk and responsive regulation, and the issue of regulation to foster innovation has enjoyed less attention. Indeed, concerns about fostering innovation through regulation date from at least 1995 when the European Commission issued a *Green Paper on Innovation*,[3] stressing that 'fostering a legal and regulatory environment friendly to innovation' was one of its core objectives. This was followed a year later by the *First Action Plan for Innovation in Europe*, and more recently with the *EU's Europe 2020 Strategy*, designed to foster an innovation culture and create a more innovation-friendly environment. However, such statements have attracted little general academic response. An exception may be found in a paper by Kuhlman et al. who, at the end of the last century, pointed to the essential role of an innovation-friendly regulatory framework in achieving economic policy goals.[4]

The issue of regulation and innovation has also been debated in various books. For instance, in 2008, Roger Brownsword in his book, *Rights, Regulation, and the Technological Revolution*,[5] presents regulatory challenges from a perspective that is wider than only risk-orientation. He claims that we are entering an era of a new technological revolution and that this will certainly challenge the existing regulatory frameworks. Therefore, it is important to select the right regulatory instruments that are able to respond to such technological dynamics and to broader societal values. The debate in his book focuses largely on how newly emerging technological applications contribute to regulatory changes, and less so on regulatory approaches to fostering innovation. Conversely, *Regulating Technologies: Legal Futures, Regulatory Frames and Technological Fixes*, edited by Brownsword and Yeung,[6] is closer to the perspective adopted in this book as it offers a varied and double perspective, of technology as a regulatory tool and vice versa. That book emphasises how the legal community has only recently grasped the opportunities and challenges that emerging technologies pose to their 'host communities'. It focuses its attention especially on the challenges that lie in the field of risk regulation, including the role of the precautionary principle. Opportunities provided by the new technologies are also addressed in their book, especially in terms of how technology in itself provides a regulatory tool in the sense of techno-regulation (also known as regulation through code). Our book aims to build on such existing works, so as to move the debate forward to the challenges of regulation for fostering innovation and to add a more multidisciplinary view of the subject. As such, we hope that readers will experience an opening up of new perspectives on the regulation and innovation relationship.

I.3   Our target audience

This book is addressed to a range of audiences: administration scientists, legal scholars, politico-economic scientists, students (masters level and above) in these disciplines, regulators and other decision makers, plus advisors in the field of regulation and innovation. As to its academic relevance, we hope that universities and affiliated organisations will find, within their programmes in the field of regulatory affairs, that the use of this volume can support the exchange of views on balancing risks and opportunities in technological innovation – especially with regards to the discourse on enhancing, fostering and securing technological innovation in an increasingly competitive and knowledge-based world.

I.4 Format

A deliberate choice was made to present concise contributions on the regulation and innovation discourse: each contribution being of just sufficient length to offer a descriptive and analytical 'stage setting', as well as prescriptive elements for further debate. As the below overview will show, the contributions have been clustered into three parts to provide a coherent structure and a variety of leading perspectives for such debates.

The first part (of four chapters) provides a general analysis of the relationship between regulation and innovation. Subsequently, the second part of the book (again four chapters) brings forward issues related to technological innovation; providing evidence from specific regulatory areas. Finally, the third part (three chapters) aims at furthering the debate on technology innovation and regulation by focusing on the potential of newly emerging technologies to foster technology innovation.

It did not seem fitting to force some synthesis onto the contributions, or upon their groupings in parts, by adding a separate, concluding in-depth analytical contribution. Instead, the book ends with some general notes primarily intended as 'food for further thought'.

I.5 Overview of the book contents

Part I

Following this Introduction, Part I provides a general analysis of the relationship between regulation and innovation. To begin with, in Chapter 1, Antoni Brack argues that competition law has the potential to foster technology development, but that regulatory effectiveness could be better achieved through fewer regulations, thus reducing the repetitiveness and complexity of technology regulation. The next three chapters explore the main challenges facing the regulation of technology innovation in recent decades. In particular, the main concern of the authors is how to provide a legitimate, valid and effective regulatory framework that gives voice to all stakeholders involved with the technology innovation process. Contributors draw lessons from various disciplines: Donnelly draws on knowledge from classical and institutional economics (Chapter 2); Heldeweg shows the relevance of the concept of 'smart rules and regimes' in attempting to project lines along which the legal design of such smart rules and regimes could proceed (Chapter 3); and Wessel explores the role of informal international law making in the crafting of rules for technological innovation (Chapter 4).

The following paragraphs provide more detailed explanation of the chapters in this part; but readers who feel they have enough knowledge at this point may prefer to jump to Part II.

The first part of the book starts with Brack's contribution (Chapter 1). According to Brack, innovation refers to technological improvements in trade and industry that result from improvements to products, processes and organisations. Competition law is crucial, as the main factor in the development of new or improved technologies and, more importantly, for removing possible obstacles to innovative developments. In Brack's view, innovation could be enhanced by making a distinction between horizontal and vertical cartels, determining whether the undertakings involved operate on the same level or not in the production and distribution chain. In his contribution, Brack compares two separate regulations, which both allowed undertakings to cooperate, under certain conditions, for innovative purposes, but that expired at the end of 2010. The European Commission called for public consultations on the review of the expiring regime as part of the process of preparing to renew or replace this regime. Responding to this request, Brack conducts a comparative analysis of the two 'Block Exemption Regulations', and argues that these regulations should be merged into one. A reduction in the regulatory burden is an important goal of the so-called Better Regulation Programme and a simplification of these regulations would fit with this by reducing the regulatory burden on undertakings and cutting the number of definitions.

In Chapter 2, Donnelly analyses how regulation contributes to innovation generally, drawing on knowledge from classical and institutional economics. In this chapter, technological innovation is analysed from the contribution that regulation can make to economic development. As such, various instruments that, it is claimed, will enable regulation to foster competitiveness and innovation are put forward; these most frequently address standardisation, intellectual property and economic monopolies. In Donnelly's view, regulation affects the capacity of companies to attract, hold on to and utilise factors of production to protect intellectual property and to collaborate with others. He states that, very often, regulation, and the associated modes for fostering competition and innovation, provides an asymmetric balance among the various interests since regulatory rules often favour some forms of regulatory environment or economic activities over others. Regulatory bodies also bear a responsibility for ensuring that markets are kept open. This not only means combating private attempts to shut down markets, but also attempts at regulatory capture by private interests. As such, regulation is a building block that influences the capacity of companies to attract appropriate investments, including fixed and human capitals, for innovation. However, the greatest challenge of the twenty-first century, it is argued, will be developing regulatory forms that work well in supporting two different kinds of innovation within a single country: one that requires very flexible contracts and another that requires very stable ones.

In Chapter 3, Heldeweg argues that 'smart rules and regimes' form the main building blocks in fostering technological innovation without neglecting safeguards against technological risks. Adequate regulation is crucial for enhancing technological innovation. Supporting the arguments of the Dutch Scientific Council for Government Policy (WRR) study, Heldeweg shows that innovation is indeed a complex system, focusing not only on the creation of new knowledge and technologies but also on changes in organisation, management and labour, and on the diffusion and application of new knowledge. As such, providing an effective legal design methodology is crucial. Given this relevance, this chapter attempts to suggest lines along which the legal design of such smart rules and regimes could be envisaged. Heldeweg argues that the smartness of rules and regimes relates to balancing two variables: high innovation dynamics and strong conflicts of interest. It is thus crucial that the legal norms and regulatory frameworks align with the current state of a technology, and this could raise issues of legitimacy and legal validity. To this end, the chapter explores an approach based on institutional legal theory. This offers a significant first step but a legal design methodology for 'smart rules and regimes' requires further steps to provide a science-based perspective on which innovation regulation can be built. Heldeweg concludes his contribution by arguing that technology innovation could be fostered through 'smart rules and regimes'. These regimes would need to be legitimate (delineating between state power and citizens' freedom), have legal validity and be effective and efficient. In this respect, hybricity among the various regulatory actors will be crucial in providing balance among the multiple interests across the various modes of coordination (hierarchy (government); competition (markets) and collaboration (social networks)).

The contribution by Wessel in Chapter 4 advances the idea of regulating technological innovation through informal international law. Wessel indicates that the role of international organisations in regulating technological innovation has been crucial in several sectors, including telecommunications, health and safety, and intellectual property. Such organisations are generally involved in normative processes that, de jure and de facto impact on states and even on businesses and individuals. As such, in *Wessel's* view, the international regulation of technology seems to have been taken over by non-traditional international bodies. In fact, he argues that it would even be fair to say that global governance in the area of technology is no longer directly in the hands of the traditional international actors, the nation states. A variety of governmental, non-governmental and hybrid organisations are today involved in the crafting of rules and standards for worldwide application. This chapter puts forward the view that 'informal international law making' has indeed become a tool to regulate technological innovation. On the one hand, this has brought a new mechanism to the regulation of technology innovation but, on the other, there are many disputes concerning the legitimacy and accountability of such rules. By looking at a number of 'informal' bodies involved in the regulation of the Internet, an attempt is made to answer the question as to the extent to which the activities of international non-governmental actors can nevertheless be seen as an expression of international public authority.

Part II

The second part of the book brings forward issues related to technological innovation, providing evidence from specific regulatory areas. The first two contributions (Chapters 5 and 6) focus on the telecommunications sector. Here, the roles/potentials of the emerging networks (Broos, Chapter 5) and competition law (Holterman, Chapter 6) in fostering innovation in this regulatory field are discussed. Both of these contributions advance the idea that ensuring certainty and incentives is crucial if the telecommunications sector is to innovate. Sanders and Luisa Marin then focus on other regulatory fields. In Chapter 7 is Marin's work on the use of technology innovation in policing the external EU borders. This part concludes with Chapter 8, where Sanders argues for safeguarding public interests in the innovation process, and ensuring legitimacy and effectiveness by drawing lessons from the energy sector.

As before, for those who want a little more information on the content of these Part II chapters, we provide a more detailed description of each contribution below. Other readers may be satisfied with the above brief synopsis and wish to move on to Part III.

The contents of the chapters in this part are organised as follows. In Chapter 5, Broos advances from the limited scientific insights that currently exist into the relationship between regulation and innovation, by providing a new perspective on the possibilities, or otherwise, for pacing innovation through regulation. In particular, he focuses on the meso-level (industry) discussions about stimulating the deployment of so-called Next Generation Networks (NGNs) for telecommunications. Broos argues that the structured application of innovation timing theory, combined with the concept of network effects in the telecommunications industry, and related to new entrance strategies provides a better understanding of the influences of several regulatory measures because it exposes the network effects that enhance some first-mover advantages. In an emerging telecommunications market, a first mover is likely to become a provider with sustainable market power. If a leapfrog-enabling technology gateway becomes available, investing in an NGN appears to be a more attractive strategy for new entrants in an unregulated telecommunications market than investing in a 'Same Generation Network' or attempting a service-based competition. By confronting several common regulatory practices in telecommunications with innovation timing advantages and network effects, Broos analyses the institutional influence of these practices on the attractiveness of the strategic options open to new entrants. The analysis shows that cost-based mandated access leads to a declining attractiveness of investing in NGNs, whereas relaxing mandated access obligations appears to influence the development of NGNs positively. Furthermore, this analysis leads to a better explanation of why interconnectivity and portability obligations, as well as the guaranteeing of greater regulatory certainty, have positive effects on all new entrance strategies.

In Chapter 6, Holterman again analyses the telecommunication industry, this time providing a positive and normative discussion on the role of competition law and competition authorities in fostering, or hampering, innovation in this regulatory field. Holterman develops his argument, as to whether it would be appropriate for competition authorities to interfere in the DSL industry (concerning particular forms of digital data communication, such as ASDL and VDSL), by discussing the Deutsche Telekom and Pacific Bell cases as examples of competition law 'intruding' into a regulated industry. He combines an analysis of these recent EU and American case law examples with a simple economic model of multi-agency regulation and its effects on innovation. Most importantly, Holterman's contribution sheds light on the 2002 EU telecommunications package, as well as the choices made by the various actors in the Deutsche Telekom situation to determine what role the EU legislators actually intended for the competition authorities. While the decisions of lawmakers should ideally reflect their careful consideration of the consequences of their decisions on innovation, Holterman concludes that the 2002 EU telecommunications package appears to be highly problematic in that it seems to give competition law an enhanced role in discouraging future innovation. According to Holterman, this development has created many doubts because regulators are normally in a much better position to understand the market, since competition authorities are only able to intervene ex post, and this has created great uncertainty among market actors, reducing their willingness to invest in innovation.

The analysis by Marin in Chapter 7 focuses on a different regulatory field in which innovation and technology are also crucial. She provides interesting evidence on the use of technology innovation in the policing of the external borders of the European Union, focusing primarily on two technology initiatives: the European Borders Agency (FRONTEX) and the European Border Surveillance System (EUROSUR). Starting with the origins, tasks and responsibilities of the EU agency FRONTEX, as well as the FRONTEX-led operations carried out on the southern maritime border, Marin argues that the policing of external borders emphasises elements of multilevel governance that involve various stakeholders and interests. In the second part of her contribution, Marin analyses the various technologies and systems that are in place for border control, and examines in particular the EUROSUR, which offers a good example of how technology can be exploited in the context of managing external maritime borders. This chapter concludes by discussing the political and legal implications of these technology initiatives, and makes recommendations for their future improvement.

In Chapter 8, Sanders analyses how one can safeguard the public interests in the energy sector. His empirical analysis starts with the Dutch government's plans to increase energy supply from renewable energy sources by negotiating climate agreements on national and international levels. Realising that the energy sector's sustainability ambition is dependent on technological innovation, the government has launched several projects in collaboration with private parties. An example of such an initiative is the Salland Green Gas Project. Based on an analysis of this project, Sanders illustrates that the social interests, which the government designates as public, are in fact variable and that this dynamic spills over into the governance structures in which public interests are embedded. Following this, he analyses how the government can ensure the functioning of the Salland Green Gas Project while balancing the legal and public administration values of effectiveness and legitimacy.

Part III

The third part of the book has three chapters. In this part, the contributors further the debate on technology innovation and regulation by focusing on the potential of newly emerging technologies to foster technology innovation. In the first chapter (Chapter 9), Chowdhury discusses the field of nanotechnology, a newly emerging technology, through a comparative analysis of the various sub-political actors that engage in regulating nanotechnology without any de jure legal mandate so to do. In the last two chapters (Chapters 10 and 11), Kica and Rodriguez respectively focus on the role of patents in spurring innovation and economic growth, and discuss the current challenges facing the European patent system in coping with the pace of technology development and new innovation requirements.

Again, for those readers who desire more knowledge on the content of the chapters in this third part at this stage, the following paragraphs offer brief summaries.

Following the debate on emerging technologies, Chowdhury (Chapter 9) sheds light on the participation of various actors in the development of regulatory norms within the field of nanotechnology. In her chapter, she revisits the arguments provided by theorists for and against national versus international regulation of nanotechnology. As a newly emerging technology, nanotechnology faces many challenges since there is a lack of information as to what regulatory actions this field could be based upon, as well as on the toxicity, and health and safety aspects of nanomaterials. Chowdhury emphasises the reality that several international actors have become active in setting up expert groups and other administrative structures in order to develop regulatory norms in the context of nanotechnology. Interestingly, some of these actors have had international exposure and experience in developing regulations in other new and emerging technology areas. For others, this is an opportune area for extending their regulatory competencies. However, such actors share a common characteristic: they are essentially sub-political in nature. This concept of sub-politics has been developed by the sociologist Ulrich Beck to characterise international actors that engage in regulatory norm setting without having any de jure legal mandate to do so. This chapter explores in detail three such actors within the domain of nanotechnology regulation: the IFCS (Intergovernmental Forum on Chemical Safety), the OECD (Organisation for Economic Cooperation and Development) and the IRGC (International Risk Governance Council). Linkages between these three actors and EU institutions exist at numerous levels through membership, common normative foundations and functional linkages that lead to converging interests. Further, given the growing acceptance of international forums/institutions as efficient and effective sites for regime creation, there is a high probability that the norms emanating from such sub-political actors will seep into the EU legal regime through the pathways identified above. In the final part of her contribution, Chowdhury analyses the regulatory partnerships between EU institutions and such sub-political actors in terms of delegation and exchange of competences with reference to norm creation for EU nanotechnology regulation. She concludes that there have been several structural innovations which have been adopted by EU policymakers, including dividing, delegating and sharing specific competences within highly technical domain areas such as nanotechnology, but that there remain questions on the transparency and legitimacy of this policy formulation process.

In Chapter 10, Kica starts by emphasising the role of patents within the innovation process and technology development. Following this, she focuses on the challenges that the emerging new technologies and the increase in the number of patent applications have brought to the IP regulatory framework and the ability of patent examiners to issue high-quality patents. The quality of the granted patents is considered to be the main endogenous factor challenging the ability of the patent system to encourage innovation and the diffusion of technology. Patent quality in this chapter is viewed as the extent to which patents meet patentability standards, an aspect which is often assessed by patent examiners. As such, Kica argues, the quality of patents depends on the competence of the examiners as well as on the time and search materials available to them. However, within the current European patent system, there are too few patent examiners which, coupled with the shortage of time available to search and examine patent applications and the rapid developments in emerging fields of science, has increased uncertainty in the interpretation of patentability criteria and in granting patents to potentially valuable inventions. This chapter acknowledges the complexity of the patent system and examines the potential of administrative mechanisms that have been put forward to improve patent quality. It starts by examining the scholarly debate on the mechanisms that support patent quality enhancement through improving the examination process. Next, it lays out the landscape of the administrative mechanisms and addresses the quality of search and examination procedures, and of the quality of products and processes at the European Patent Office (EPO). It distinguishes between the strategies that address the quality of search and examination procedures, and the EPO's use of work from other patent offices and sources. Finally, in her contribution, Kica brings the scholarly debate and the functioning of the administrative mechanisms together, and provides policy recommendations aimed at enhancing the value of patents for newly emerging technologies. She concludes that patent quality is a shared responsibility of both patent examiners and applicants, and that these need to work together to enhance the functioning of the patent system, and so foster innovation activities and technology development.

In Chapter 11, Rodriguez takes a similar line, arguing that patents have a role in fostering technological innovation in the European Union. According to Rodriguez, patent protection is crucial and various actors in many sectors invariably opt to protect their inventions while building monopoly positions or establishing a financial strength in the market or a vital position during the standardisation process. However, the current patent system in Europe is considered to provide little added value to innovation. Rodriguez argues that the current European regime allows member states to retain their institutional arrangements and prevents any moves to delegate responsibility beyond the national sphere. Such a regime can be characterised as a fragmented European patent system of national translation, validation and enforcement. Fragmentation is regarded as a failure of the system since it leads to higher costs, uncertainty and to low quality. At this point, Rodriguez argues that the most appropriate way to overcome the failures of the current EU system is through a unitary title and a centralised patent court that could enhance technological innovation at various levels (micro-, meso- and macro-). However, there is still significant conflict and debate among member states as to how to establish such a unitary patent system. Given this situation, Rodriguez posits that enhanced cooperation could be an alternative strategy, and that several countries could work together without the unanimous participation of all EU member states.

I.6   Collaborative efforts and contributors

Focused work on the chapters of this book started in September 2010 and contributions were completed in July 2011.

The contributors are all active in the fields of 'European regulation' and 'regulation in Europe', some following a more general, theoretical perspective and others focusing on a more specific material object of regulatory study, but all relating to technological dynamics or innovation. All contributors have previously written for and published their findings and opinions in international books and journals. Some authors are already leading experts in their fields (Wessel in European law, Rodriguez in patent policy, Heldeweg in environmental legal governance, *Donnelly* in global governance and the politics of economic policy, Marin in European constitutional law and Brack in competition law and technology regulation), others are involved in PhD projects in the field of regulating innovation.

The book evolved as a collective journey. The editors and contributors have collaborated in the course of meetings on Innovation and Governance Studies at the University of Twente, and the editors took it upon themselves to arrange several meetings in which the initial ideas on the focus of a book and, later, abstracts and draft contributions were discussed. Although the book covers a variety of related foci, angles and perspectives, the contributors see the book as an outcome of their collaborative efforts.

Notes

1. See Black, J., M. Lodge and M. Thatcher (eds), *Regulatory Innovation* (Cheltenham: Edward Elgar, 2005).
2. Black, J. 'Critical reflections on regulation', *Australian Journal of Legal Philosophy*, vol. 27 (2002), pp. 1–35.
3. European Commission. *Green Paper on Innovation* (Luxembourg: European Commission, 1995).
4. Kuhlman, S, C. Bättig, K. Cuhls and P. Viola, *Regulation und künftige Technikenentwicklung: Pilotstudien zu einer Regulationsvorausschau* (Heidelberg: Physica, 1998).
5. Brownsword, R. *Rights, Regulation, and the Technological Revolution* (Oxford: Oxford University Press, 2008).
6. Brownsword, R. and K. Yeung (eds), *Regulating Technologies: Legal Futures, Regulatory Frames and Technological Fixes* (Oxford and Portland: Hart Publishing, 2008).

Part I
General Analysis of the Regulation and Technological Innovation Relationship

1

Regulation *for* Innovation: A Comparative Inquiry into a Regulatory Pair of Twins

Antoni J. P. Brack

1.1   Introduction: The reason of this research

In this contribution 'innovation' refers to technological improvements in trade and industry that result from improvements of products, processes and organisations. Regulations aimed at stimulating such innovations by providing opportunities, under certain conditions, for undertakings to work together belong to competition law. This legal area is at the intersection of public administrative law and private business law.

The instrumental element of law and legislation is often overlooked. Non-normative, neutral judicial tools, such as agreements, patents and trademarks, enable management to establish relationships and protect properties. Legal persons are the building blocks for establishing a group of subsidiary companies subordinate to the management board of a shareholding corporation. The normative element of competition law is obvious: strategic management of an undertaking is severely limited in its ability to make arrangements with competitors and the long-term consolidation of undertakings in the form of a merger or takeover is, in principle, prohibited without prior consent from an antitrust authority. Within the primarily normative system of competition law, there are nevertheless some legal instruments that aim to facilitate cooperation between undertakings under certain conditions. The goal is to achieve certain societally desirable results that would have been difficult or impossible to achieve without these forms of cooperation. This is done through the use of the legal vehicle of the Block Exemption Regulation (BER). We will provide a short introduction of the main points of competition law in order to explain how cooperation between undertakings is stimulated within a system that is primarily focused on preventing this cooperation distorting competition between undertakings.

Competition law: System and policy

This contribution is not about the private law version of 'competition law', which is the domain of topics such as trade practices, commercial publicity, trademarks and advertising. Conversely, our concern here is with the public law version: the influence of the regulatory agency on the degree of competition in markets. The Anglo-American notion of *anti*trust law is a better indicator of the key issue, the prohibition of restrictive practices. Competition, that is at least non-cooperation between firms, is the norm. European law, and more specifically European competition law, has a supranational dominance. The Treaty on the European Union articulates (in Article 2) – among other objectives – the promotion of economic and social progress, in particular through the creation of an area without internal frontiers. The common rules on competition are an important means to achieving these objectives. They are the foundation of European competition law, which, as a system, basically consists of the following three parts:

1. Prohibition of Cartels (Article 101 Treaty of the Functioning of the European Union (TFEU), ex article 81 TEC);
2. Prohibition of the Abuse of a Dominant Position (Article 102 TFEU, ex article 82 TEC);
3. Control of Concentrations (EC Merger Regulation).[1]

This could be denominated as the first level of European competition law. In light of the subject of this chapter, regulation for innovation, we will focus on the first part (1) above). Cartels are anti-competitive agreements. The broadly defined term 'collusion' is often used because it covers all the categories of anti-competitive behaviour mentioned in Article 101 TFEU (see text in Annexure) namely agreements (written as well as oral), decisions of associations and concerted practices (i.e., coordination of conduct). A difference with USA antitrust law, which is directly focused on maintaining competition, is that the creation of the common market – an area without internal frontiers – is the EU's main objective.[2] It's not so much the restriction on competition, but the restriction on trade between member states that should first and foremost be prevented. Fostering good competitive relations is an alternative objective. Or perhaps even better: 'EU competition law can be seen as serving two masters, the "competition" one and (even more demanding) the imperative of single market integration'.[3]

Collusions are prohibited as incompatible with the common market if they have as their objective or effect the 'prevention, restriction or distortion' of competition within the common market. Well-known examples of these restrictive practices are the fixing of prices, the controlling of production volumes, the sharing of markets. Collusions or cartels are automatically void and can be sanctioned with very high fines. The principal rule, that cartels are prohibited, cannot always and absolutely be maintained. This would constitute a severely unsubtle approach. The rule must be applied in an efficient and effective manner. According to the so-called *de minimis* rule, insignificant cartels that do not have a noticeable effect on competition will not be dealt with. Another, more important derogation from the principal rule states that a certain level of restriction on competition is allowed, provided that important societal advantages can be achieved, and provided that competition will not be restricted any further than is necessary to achieve these advantages. At this point competition law meets the overall objective of the Treaty: economic progress. Article 101(3) TFEU provides the possibility of exempting certain agreements from the prohibition set out in Article 101(1) TFEU on the basis of policy considerations. The agreements are then no longer void, but perfectly valid. The Prohibition of Cartels may be declared inapplicable in the case of any agreement (and any decision by an association, and any concerted practice).

> which contributes to improving the production or distribution of goods or to promoting technical or economic progress, while allowing consumers a fair share of the resulting benefit.

It is obvious that this part of European competition law makes it possible to develop and maintain innovation-fostering policies.[4]

The European Commission's practice of applying Article 101(3) TFEU has undergone two important changes as the Community developed over the years. Initially, the Commission only granted individual exemptions. After gaining experience in this mode of application, the Commission started to codify these individual cases in so-called block (or 'group') exemptions, which state the conditions for the application of Article 101(3) TFEU. The procedure for companies that wanted to cooperate in one way or another now became as follows: first one would look for the relevant block exemption and assess whether the cartel meets the criteria, so that the wording could be aligned with these criteria in the initial drafting of the agreement. If there was no relevant block exemption or if it proved impossible to adjust the cartel to the available exemption, the second step would follow: the request for an individual exemption.

After European corporations and the European bureaucracy had operated this way for years, an important change came into effect as of 1 May 2004.[5] By far the most important change concerned the abolishment of the possibility for cooperating undertakings to submit their agreement to the Commission prior to obtaining dispensation. The backlog in the assessment of applications had become too extensive and the Commission wanted a new and more efficient approach. Therefore, as of 1 May 2004 only ex post assessment will take place.

Block Exemption Regulations[6]

Drafting a BER is a creation of legislation on the second level: a detailed and specific elaboration of a rule of the first level. A typical BER in the field of competition law includes the conditions under which an agreement will be allowed that is to say formally that the cartel prohibition rule is declared inapplicable in the case of agreements that fulfil the terms formulated in the BER. This allows the legislator to determine specific policy objectives for a certain area.[7]

At this point it would be adequate to introduce another basic distinction that is essential to the application of competition law. This is the distinction between horizontal and vertical cartels. The criterion for the distinction is the answer to the question whether or not the undertakings involved are operating at the same level in the production and distribution chains. Suppose that the contracting parties are both in the wholesale trade of certain spirituous beverages. Such an agreement is classified as a horizontal cartel. If, on the other hand, the parties are on different levels of the supply chain, their agreement would be classified as a vertical one, like the agreement between an importer and the dealers of a certain brand of cars. Companies that become part of a horizontal cartel are restricting competition per se, because they are competitors. Vertical agreements, on the other hand, are not necessarily restrictive of competition as the undertakings involved are not each other's direct competitors. This explains why there are and have been much more BERs for vertical agreements than for horizontal agreements. For several years now there has even been generally applicable BER for all vertical arrangements. On the other hand with regard to horizontal cartels only a few block exemptions have been established.[8] Two of these relate to the innovation of technology and knowledge and it certainly is interesting to research whether it's a coincidence that these block exemptions were created on the same date:

- BER on *specialisation* agreements,[9]
- BER on *research and development* agreements.[10]

1.2 The comparative analysis research

We will conduct a **comparative analysis of both regulations**.

Preview: Comparison of recitals

Instead of an Explanatory Memorandum, which usually, at member state level, accompanies a proposed piece of legislation on its way to Parliament and which, after enactment, is valuable in understanding and interpreting the provisions of that law, a legislative document of the European Union is preceded by a collection of considerations (which regularly begin with 'whereas') the sequence of which is indicated by numbers. These are commonly referred to as 'recitals'. The entirety of the recitals is essential to comprehend the full scope of such a regulation in all its parts, sections and articles.

At first glance the recitals of both BERs create the presumption that we are dealing with two very similar regulations. Both of the first recitals (1) mention the European Commission's competence to apply the exemption provision of the cartel prohibition to certain cartels ('certain categories of agreements, decisions and concerted practices') which have as their objective specialisation (Regulation 2658/2000, hereafter BER-SPEC) or the research and development of products or processes (Regulation 2659/2000, hereafter BER-R&D) respectively. The next recital (2) mentions that the foregoing regulation on specialisation (417/85) expires on 31 December 2000. In the case of R&D, two specific recitals have been inserted. On the one hand the reference to a special provision in the Treaty (163–2) on research and technological development in undertakings and their collaboration with research centres and universities and on the other hand on the possibility that these forms of collaboration may not qualify as cartels from a competition law point of view. Subsequently, the twin provision of recital (4) mentions that the foregoing regulation on research and development agreements (418/85) also expires on 31 December 2000.

The following five recitals – (3) and (5), (4, 5) and (6, 7), (6, 7) and (8, 9) – are completely identical.

Recitals (8) through (13) in the BER-SPEC are specifically regarding the content of specialisation, as recitals (10) through (16) in the BER-R&D are specific to the content of research and development.

The remainder of the recitals in both regulations are yet again identical, or more or less the same. Recitals (14), (15) and (16) in the one BER and recitals (17), (18) and (19) in the other BER are totally unisonant and contain standard considerations such as:

- the exemption may not cover restrictions of competition which are not indispensable for the expected benefits,
- certain so-called hard core[11] restrictions of competition, such as price fixing, are not allowed in any circumstances,
- as a final piece of regulatory responsibilities the Commission has the possibility of hitting a legal safety break: should an exempted agreement nevertheless, unexpectedly, have severe negative effects on competition, the Commission is empowered to withdraw the benefit of the block exemption for this agreement.

Recitals (17), (18) and (19) in the one BER and recitals (21), (22) and (23) in the other contain the concluding recitals, which are of a rather formalistic nature.

Comparison of articles

These two BERs replace their predecessors which both expired on 31 December 2000. The BER-SPEC and the BER-R&D have many provisions in common and more or less the same structure. In fact there is a remarkable resemblance, as we would have expected from the comparative overview of the recitals. Both regulations contain nine articles with headings. These headings are similar for seven out of the nine articles. The contents of these articles will be discussed below. First, a short overview will be given.

Article 1, titled 'Exemption', stipulates the exemption from the cartel prohibition and distinguishes between three varieties of *specialisation*, and respectively three varieties of *R&D*.

Article 2 ('Definitions') defines the key terminology.

In the BER-SPEC Article 3 has the title 'Purchasing and marketing arrangements' and contains an additional exemption for an exclusive purchase and/or supply obligation in the context of a specialisation agreement, or for a joint distribution agreement. In the BER-R&D on the other hand, Article 3 ('Conditions for exemption') stipulates additional conditions for the application of the exemption.

In addition, in the BER-SPEC Article 4, entitled 'Market share threshold', establishes a maximum market share of 20 per cent, while in the BER-R&D Article 4 ('Market share threshold and duration of exemption') the market share is established at 25 per cent. The latter article also contains special, additional provisions on the validity period of the exemption in specific cases in practice.

The other articles, in line with the first two articles, have similar headings. Article 5, 'Agreements not covered by the exemption', lists in both regulations so-called hard core agreements that are never allowed in any domain of competition law and therefore called black clauses. Both Articles 6 are titled 'Application of the market share threshold' and contain a flexible application of the market share criteria. Article 7 in both regulations is entitled 'Withdrawal', which refers to the competence of the Commission to withdraw the benefits of the BER. Article 8 ('Transitional period') formulates a rule for the transition from the preceding Regulations No. 417/85 and 418/85 to the current No. 2658/2000 and 2659/2000. And Article 9 finally, entitled 'Period of validity', states that the current regulations will come into effect on 1 January 2001 and expire on 31 December 2010.

Methodology of comparative analysis

In the light of this comparative study we can assess the provisions by applying a threefold system of qualifications with an increasing scale of substantive difference: identical, generic and specific. The qualification *identical* means that the provision in both BERs is textually identical both in content and heading. Two out of the nine articles meet this qualification, namely the last two articles of both regulations. If we apply this criteria more loosely to sections of the articles, then a much larger portion of the regulations qualifies as identical. *Generic* means that the text of the article contains a reference to specialisation agreements or R&D agreements, but the text of the comparable articles is mostly identical. Four out of nine articles are considerably generic, but also contain specific elements. An article is qualified as *specific* if the text (or part of the text) is focused on specialisation or R&D. Three out of nine articles have an important or predominantly specific content.

Provisions specific to specialisation

Article 1 BER-SPEC declares that the prohibition of cartels in the Treaty (art. 101-1) is not applicable (according to art. 101-3) to the following three types of specialisation agreements between two or more undertakings:

- *Unilateral specialisation* agreements: one party agrees to cease or refrain from the production of certain products and purchase them from a competitor while this competitor agrees to produce and supply those products (party A no longer produces product X and agrees to purchase this product from party B).
- *Reciprocal specialisation* agreements: two or more parties agree to cease or refrain from producing certain different products mutually and

purchase these products from the other parties (party A no longer produces product X and agrees to purchase this product from party B while party B no longer produces product Y and agrees to purchase this product from party A).

- *Joint production* agreements: two or more parties agree to produce certain products jointly, that is together, collectively.

The only point of exempting an agreement from the cartel prohibition is if otherwise the prohibition would apply. In particular, if parties to the agreement are not competitors – for instance because they are active in different markets – the cartel prohibition will not apply and an exemption will not be necessary. This also goes for provisions contained in specialisation agreements that are directly related to the primary object of the agreement, such as IP rights. One can imagine that, for instance, in a reciprocal specialisation agreement the mutual licensing or 'cross licensing' of these rights is a necessary additional provision.

Article 3 BER-SPEC widens the exemption to cases where the parties accept an exclusive purchase and/or supply obligation in the context of the three types of specialisation arrangements. There is also an expansion of the exemption foreseen in the case that parties agree not to sell the products resulting from their specialisation efforts themselves independently but organise joint distribution instead, or in the case of joint production under certain conditions, provide for a third-party distributor.

Article 4 BER-SPEC contains the important condition for exemption that the combined, that is collective, market share of the participating enterprises including connected companies does not exceed 20 per cent of the relevant market, which is the market for the products that are the result of specialisation.[12]

Article 5 BER-SPEC ('Agreements not covered by the exemption') consists of two sections and states in the first that the exemption is not applicable to certain types of agreements. The second section provides for an exception. It is a given fact in competition law that certain severe impediments on competition can never be allowed, and can thus never be exempted from the prohibition. Cartels that contain these kinds of impediments are called hard core, although the impediments themselves are also referred to by the same term. In general, as well as in this case, this refers to price fixing, the limitation of output or sales and the allocation of markets or customers. These restrictions on competition are unacceptable because they respectively eliminate price competition, have a price enhancing effect under a similar demand and distort the unrestricted allocation of demand and supply. Surprisingly, in specialisation cartels these restrictions between participants are allowed in certain circumstances. In a unilateral or reciprocal specialisation agreement arrangements can be made on production volume. This is also allowed in regard to a joint production agreement. In the case of joint distribution, arrangements are allowed on sales targets and prices charged to third parties.

Article 7 BER-SPEC is concerned with the competence of the Commission to withdraw the exemption in certain cases. This article is formulated in very general terms: if the exemption of a certain cartel based on this BER means that unforeseeable effects occur which are incompatible with the conditions of Article 101(3) Treaty, the Commission can withdraw the advantages of this exemption from this cartel. This provision follows by stating that this applies particularly in two situations, although very general terms are applied here as well, namely that the agreement does not bring about significant rationalisation results or if consumers do not receive a fair share of the benefits.

Provisions specific to R&D

The central definition in the BER-R&D is 'the acquisition of know-how relating to products or processes and the carrying out of theoretical analysis, systematic study or experimentation, including experimental production, technical testing of products or processes, the establishment of the necessary facilities and the obtaining of intellectual property rights for the results'. This is a nice but rather all-encompassing description if one realises that know-how is elaborately defined as 'a package of non-patented practical information, resulting from experience and testing, which is secret, substantial and identified: in this context, "secret" means that the know-how is not generally known or easily accessible; "substantial" means that the know-how includes information which is indispensable for the manufacture of the contract products or the application of the contract processes; "identified" means that the know-how is described in a sufficiently comprehensive manner so as to make it possible to verify that it fulfils the criteria of secrecy and substantiality'.

This definition clearly leaves nothing to chance, though it is questionable whether it really needs to be so elaborate.

Article 1 BER-R&D declares that the prohibition of cartels in the Treaty (art. 101-1) is not applicable (according to art. 101-3) to the following three types of agreements:

- *joint R&D* of products or processes *and joint exploitation of the results;*
- *joint exploitation of the results* of R&D of products or processes jointly carried out based on a prior agreement between the same parties;

- *joint R&D* of products or processes *excluding* joint exploitation of the results.

Because the BER-R&D applies to joint R&D and/or joint exploitation of results, the following description is useful for a good understanding of this regulation: 'research and development, or exploitation of the results, are carried out "jointly" where the work involved is:

1. carried out by a joint team, organisation or undertaking;
2. jointly entrusted to a third party; or
3. allocated between the parties by way of *specialisation* in research, development, production or distribution' (Article 2 section 4 under 11 BER-R&D). It is quite noteworthy that this quoted text connects aspects of research and development with specialisation and thus emphasises the *interrelationship of the two BERs*.

Article 3 BER-R&D is a substantially different provision than Article 3 BER-SPEC. The latter could just as easily have been included in Article 1, since it contains an extension of the exemption. Article 3 BER-R&D ('Conditions for exemption'), however, states that exemption depends on the following specific conditions:

- for the purpose of further research or exploitation all the agreeing parties, cartel insiders, must have access to the results of the joint R&D[13],
- each party must be free to exploit the results of the joint R&D independently although such exploitation right may be limited to some technical fields of application where the parties in their pre-contractual period were not competing,
- any joint exploitation must relate to results protected by IP rights or constitute know-how, which substantially contributes to technical or economic progress and must be decisive for the manufacture of the contract products or the application of processes.

Finally, in the following condition there is an *overlap* with the block exemption regulation on specialisation: undertakings that fall under a R&D agreement and are responsible for production as a form of specialisation must accept and fill orders from all parties, unless the R&D agreement also provides in joint distribution. This last condition can be understood in a way that, in the case of joint distribution, not all the parties have to be supplied, but merely the joint distributor.

At first glance, Article 4 BER-R&D, entitled 'Market share threshold and duration of exemption', seems much more complicated than the simple Article 4 BER-SPEC which just formulates the condition of the market share maximum. However, on closer inspection the text appears to be needlessly complicated. This is caused by the text of the provision that distinguishes between two situations: on the one hand, parties are either competitors or not, and on the other hand it contains a much too complicated formulation of the period of exemption.

If the parties are *not* competitors, the duration of the exemption will be as long as is required for R&D. It is questionable whether this provision is useful, since cooperation will not constitute a restriction on competition if the parties aren't competitors. The duration of joint R&D is also not subject to any time restrictions. If the R&D results are jointly exploited, the exemption will continue for seven years, from the moment that the jointly developed products are first introduced to the common European market. This period appears to be arbitrary and because no other indication can be found, we will assume that the biblical seven fat and seven lean years were the legislator's source of inspiration.

When two or more of the participating undertakings are indeed each other's competitors, the above-mentioned period[14] of exemption will only apply when the combined market share of the participating undertakings does not exceed 25 per cent of the relevant products. It is remarkable though that the moment determined to measure the market share is the moment that the parties joined the R&D agreement. Apparently, it does not matter what happens to this market share later on. Finally, the article states that after the initial period of exemption has expired, the exemption will remain in effect as long as the combined market share does not exceed 25 per cent. Thus, it is possible that during the initial seven fat years the combined market share exceeds the initially allowed maximum significantly, without any implications from a competition law point of view. It seems much easier and more effective to begin linking the exemption to the maximum allowed market share at the start of the R&D cooperation.

Article 5 BER-R&D contains 'Agreements not covered by the exemption' and is more comprehensive than its equivalent in the BER-SPEC. The article contains a long list of prohibited agreements, that is to say agreements that cannot fall under the exemption and are therefore likely to be affected by the cartel prohibition of the Treaty. Paragraph 2 contains two exceptions to this rule, so that in these situations the exemption does, yet again, apply.

The 'agreements not covered' include, first of all, the previously mentioned 'hard core' agreements, such as limitation of output or sales and price fixing, that are forbidden throughout the domain of competition law. The other agreements not covered are somehow related to R&D arrangements and therefore qualified as specific in this paper. Most of these agreements contain restrictions that go beyond what is strictly necessary to achieve the goal, namely joint R&D and/or joint exploitation of results. An example is the limitation of the parties' freedom to conduct independent research, or in cooperation with third parties, in an area that is not related to the current agreement. Another example is the protection of the developed knowledge with IP rights for a much longer period than is necessary for the R&D cooperation. Some prohibitions are related to the already mentioned period of seven years: if it benefits successful R&D the number of clients that can be approached may be limited for this period of time. In the same way, the parties can limit the areas in which an active sales policy can be conducted.

The 'Withdrawal' Article 7 in the R&D regulation is more extensive than in the other block exemption regulation. Most reasons are specifically related to R&D, namely:

- the existence of this R&D agreement restricts third parties substantially in carrying out R&D because of limited research capacity elsewhere,
- the existence of this R&D agreement restricts third parties substantially in accessing the market for the contract products because of the particular structure of supply,
- the existence of this R&D agreement makes effective competition in R&D on a specific market impossible,
- without any objectively valid reason the parties are not exploiting the results of the joint R&D.

It is remarkable that the apparently different reasons for a limited or absent R&D rest capacity and the further impossibility of R&D competition seem to come down to essentially the same thing. It is crucial for the exemption in competition law that there is a possibility to withdraw the benefits of the exemption if it becomes apparent later on that without an objective reason it is impossible to create results of societal or economic relevance.

Remaining non-specific provisions: Generic and identical

Article 2 BER-SPEC contains a number of definitions that are relevant for a good understanding, correct interpretation and relevant scope of this regulation. Some are merely formal such as the definition of 'agreement', which repeats part of the text of Article 101 TFEU, which describes cartels as 'agreement, a decision of an association of undertakings or a concerted practice'. Needless to say that the text 'agreement means an agreement ...' cannot qualify as a definition. One can also question whether it is necessary to define 'product' and 'production', although these terms are given a somewhat broader definition than usual: the provision of a service is included. Other definitions are also not specifically related to innovation improvement, but are more or less customary in the area of competition law, such as the term 'connected undertakings', which, put simply, refers to undertakings that are connected to each other in a group structure. 'Participating undertakings' means undertakings party to the agreement and their respective connected undertakings. This unexpectedly broad definition is of major importance to the market share limit: undertakings connected through a group structure profit a lot less from these block exemptions than independent, or so to say stand-alone undertakings.

A similar term is 'competing undertaking' which means not only an actual competitor but also covers a potential competitor: 'an undertaking that would, on realistic grounds, undertake the necessary additional investments [...] so that it could enter the relevant market'. Many of these definitions are identical and rather redundant because they correspond to normal spoken language or have a more or less standardised meaning within the field of competition law. The definitions that are not specifically related to R&D are *identical* to the definitions of these terms in the other regulation.

As mentioned before, in both Articles 5 ('Agreements not covered by the exemption') the definitions of 'hard core' cartels and 'hard core' clauses in agreements are *identical.*

The BER-SPEC applies a market share of 20 per cent of the relevant market, whereas the BER-R&D applies a percentage of 25. In both Articles 6 ('Application of the market share threshold') two subjects are regulated. It is indicated how the percentages should be calculated, namely preferably on the basis of market sales value or, where this is not possible, on market share volumes. Furthermore, the market share rule is softened, so as to prevent fluctuations in market shares that would result in the block exemption sometimes applying and sometimes not. This is a matter of effectiveness and legal certainty. If the increase in the market share is limited to a maximum of 5 per cent, the exemption will remain in effect for a further two years. If the increase exceeds 5 per cent, the exemption will remain in effect for a year. An undertaking cannot receive a suspension for more than two years in total.[15] Except for the percentages, the text of the articles is *identical*. The two final provisions of both regulations are completely *identical*. Article 8 ('Transitional period') played an important role when both regulations came into effect on 1 January 2001. Articles 9 ('Period of validity') declare that the regulation shall expire on 31 December 2010.

1.3   Evaluation and recommendations

The existence of competition law as a system of rules concerning collusions and limitations of competition between undertakings is widely known within business enterprise circles; cartels are prohibited. It is therefore, at least in a systematic sense, nevertheless remarkable that collaboration between undertakings is allowed in case of business agreements 'which contributes to improving the production or distribution of goods or to promoting technical or economic progress' (Article 101(3) Treaty). Specialisation in production and increase of know-how related to products and processes through R&D are supposed to be such contributions. This sends an important positive signal to the business community.

First of all, because they do not have an influence on interstate trade or because the cooperating businesses are non-competitors or do otherwise not distort competitive circumstances, many business agreements on cooperation for innovation and technology development will probably not even qualify for the prohibition rule of Article 101 Treaty. And even if they do cause such an anti-competitive effect, their collective market share might not exceed the allowed maximum and stay within the so-called safe harbour zone.

Even though there is very little insight into how much undertakings gain from the advantages of these block exemption regulations, and even though it is not certain which specialisation and R&D agreements would fall under the cartel prohibition without these BERs, it is – as such – a positive point that these BERs emphasise that collaboration between undertakings is allowed in the light of innovation and technical and economic progress.

Because the block exemption regulations on specialisation and R&D appeared very similar at first glance, this comparative study was carried out. The number of articles and parts of articles in BER-SPEC and BER-R&D that are identical or generic exceeds the number of specific provisions. This makes it worthwhile to assess whether both regulations can be brought together into one. Reducing the number of regulations would be an important contribution to lowering the regulatory burden on business enterprises. One could consider including the BER on Technology Transfer[16] in a follow-up study, because it aims to achieve the same policy objectives.[17]

Given that the substance of these regulations is regarded as positive, the question arises, not if but in what form they will be continued after 31 December 2010. Preparing an answer to this question we can connect to a part of the text in the identical recital (3 BER-SPEC and 5 BER-R&D):

> A new regulation should meet the two requirements of ensuring effective protection of competition and providing adequate legal security for undertakings. The pursuit of these objectives should take account of the need to simplify administrative supervision and the legislative framework to as great an extent as possible.

This text, which almost ten years ago was going to be part of the two Commission Regulations that were assessed in this paper, is still widely supported. Even more so, reduction of the regulatory burden is an important goal in the so-called Better Regulation Program.[18] Simplification of these regulations fits perfectly into this programme. Integration into one regulation is certainly possible and very preferable.

A further recommendation in this light is to consider reducing the number of required definitions; one cannot escape the impression that some definitions are very obvious and superficial and could be left out. After the European Commission called for a public consultation on the review of the current regime, 22 replies were posted on its website.[19] The vast majority of respondents, not surprisingly, are organisations of trade and industry and business consultants, such as law firms. Many replies showed a general request for clear-cut definitions of key concepts. Obviously, one is in need of more legal certainty under the current 'self-assessment regime', explained earlier in this paper.

Implementing another recommendation that would simplify the regulation even further would be to harmonise the market share criterion: couldn't the specialisation agreements also be exempt up to a market share of 25 per cent? At this point in time there is no good reason for these different market share criteria. It seems as though an arbitrary choice was made back then. Obviously, not without a good reason, the best half of respondents strongly urged the Commission to reconsider what is called the safe harbour threshold in both regulations.

The main recommendation to the European Commission in the consultative version of this paper was that the expiring BER-SPEC and BER-R&D be merged in one new Block Exemption Regulation on Cooperation for Innovation. The occasion of expiration and renewal creates an excellent opportunity to contribute to the support of innovation by 'Better Regulation' and reduce the legislative framework at the same time.

1.4   Concluding remarks

With hindsight, frankly, it is disappointing that even the main recommendation of this research paper to join together the two expiring regulations has been disregarded by the European Commission. Commission Regulation (EC) No. 2658/2000 of 29 November 2000 (BER-SPEC) and Commission Regulation (EC) No. 2659/2000 of 29 November 2000 (BER-R&D) have expired and were recently replaced by Commission Regulation (EU) No. 1217/2010 of 14 December 2010 (*new* BER-R&D) and Commission Regulation (EU) No. 1218/2010 of 14 December 2010 (*new* BER-SPEC)[20] both expiring 31 December 2022. The contents of the new regulations do not show any necessity or other immediate cause to continue with two separate arrangements. There are a few changes on a detailed level. There are also cosmetic modifications, presumably intended to pretend modernisations, for example, the expired regulations presented the exemption right in the first article while, for no urgent reasons at all, the new regulations start with a collection of definitions. The one and only substantial change is a loosening of the allowed exchange of information between cooperating firms, which without this permission might have been considered anti-competitive. In the recitals of both new regulations, as if it were a mantra, the just-cited consideration about the need to limit the administrative and legal burden on business enterprise is bluntly repeated: 'take account of the need to simplify administrative supervision and the legislative framework to as great an extent as possible'. It appears to be all idle gossip.

Nevertheless, the European Commission has recently stated: 'The better regulation agenda has already led to a significant change in how the Commission makes policy and propose to regulate. [...] The Commission has simplified much existing legislation and has made significant progress in reducing administrative burdens. The Commission believes that it is now time to step up a gear. Better regulation must become smart regulation and be further embedded in the Commission's working culture'.[21] Let's say that it is certainly not smart to miss a chance like this. At least in this case the Commission does not practice what it preaches.

Annexure

THE TREATY ON THE FUNCTIONING OF THE EUROPEAN UNION
Article 101 (ex. Article 81 TEC)

1. The following shall be prohibited as incompatible with the internal market: all agreements between undertakings, decisions by associations of undertakings and concerted practices which may affect trade between Member States and which have as their object or effect the prevention, restriction or distortion of competition within the internal market, and in particular those which:

   (a) directly or indirectly fix purchase or selling prices or any other trading conditions;
   (b) limit or control production, markets, technical development, or investment;
   (c) share markets or sources of supply;
   (d) apply dissimilar conditions to equivalent transactions with other trading parties, thereby placing them at a competitive disadvantage;
   (e) make the conclusion of contracts subject to acceptance by the other parties of supplementary obligations which, by their nature or according to commercial usage, have no connection with the subject of such contracts.

2. Any agreements or decisions prohibited pursuant to this Article shall be automatically void.
3. The provisions of paragraph 1 may, however, be declared inapplicable in the case of:

   • any agreement or category of agreements between undertakings,
   • any decision or category of decisions by associations of undertakings,
   • any concerted practice or category of concerted practices,

   which contributes to improving the production or distribution of goods or to promoting technical or economic progress, while allowing consumers a fair share of the resulting benefit, and which does not:

   (a) impose on the undertakings concerned restrictions which are not indispensable to the attainment of these objectives;

(b) afford such undertakings the possibility of eliminating competition in respect of a substantial part of the products in question.

Notes

1. Regulation 139/2004/EC of the Council of 20 January 2004 on the control of concentrations between undertakings, OJ (2004) L24/1.
2. For some interesting features of the Common Market *see* Donnelly, S., *The Regimes of European Integration – Constructing Governance of the Single Market* (Oxford: Oxford University Press, 2010).
3. Jones, A. and B. Sufrin, *EU Competition Law*, 4th edn (Oxford: Oxford University Press, 2011), p. 42.
4. Lugard, H. H. P. and P. M. A. L. Plompen, 'Innovatie(f) mededingingsbeleid?', *SEW* 11, 2003, pp. 372–82.
5. Council Regulation 1/2003/EC of 16 December 2002 on the implementation of the rules on competition laid down in Articles 101 and 102 of the Treaty, OJ 2003 L1/1; see Korah, V., *An Introductory Guide to EC Competition Law and Practice*, 9th edn (Portland OR, USA: Hart Publishing, 2007), Ch. 7, p. 241 ff. See also McGowan, L., 'Europeanization unleashed and rebounding: assessing the modernization of EU cartel policy', *Journal of European Public Policy*, vol. 12 (2005), pp. 986–1004.
6. The abbreviation BER is not unusual at all.
7. On the advantages and disadvantages of group exemptions *see* Korah, 99–100; see also Faull, J. and A. Nikpay, *The EC Law of Competition*, 2nd edn (New York: Oxford University Press, NY, 2007), p. 299 ff.
8. Ritter, L. and W. David Braun, *European Competition Law: A Practitioner's Guide*, 3rd edn (The Hague, The Netherlands: Kluwer Law International, 2004), p. 200 ff.
9. Commission Regulation (EC) No. 2658/2000 of 29 November 2000 on the application of Article 101(3) of the Treaty to categories of specialisation agreements, OJ 2000 L 304/3; see Ritter & Braun, 202–10; Faull & Nikpay, 701 ff; Bossche, A. Van den, *Horizontale Overeenkomsten en EG-Mededingingsrecht* (Brussels: De Boeck & Larcier, 2002), pp. 67–82.
10. Commission Regulation (EC) No. 2659/2000 of 29 November 2000 on the application of Article 81(3) Treaty to categories of research and development agreements, OJ 2000 L 304/7; *see* Ritter & Braun, 210–24; Van den Bossche, 48–66; Faull & Nikpay, 685 ff.
11. 'Hard core' is well-known jargon within the domain of Competition Law for severe restrictions of competition in any case inexcusable and incompatible with the business enterprise system. See for instance Korah, Ch. 8, 287 and further on.
12. As we will see later on there is a generic provision in both regulations providing for a flexible application of this market share condition.
13. Although research institutes, academic bodies or undertakings which supply R&D as a commercial service without normally taking part in the exploitation of the results may confine their use of the results just to further research.
14. It is assumed that this period refers to the seven years and not the period of the duration of the R&D, but this is not certain.

15. In fact a choice has to be made then between either disposing of market share or forming a yet unofficial self-judgement regarding the validity of a direct referral to Article 101(3) Treaty.
16. Commission Regulation (EC) No. 772/2004 of 27 April 2004 on the application of Article 81(3) Treaty to categories of technology transfer agreements, OJ 27.4.2004, L 123/11.
17. Lugard, H. H. P. and P. M. A. L. Plompen, 'Innovatie(f) mededingingsbeleid?', *SEW* 11, 2003, pp. 372–382. In general for instance: S. D. Anderman, *EC Competition Law and Intellectual Property Rights – The Regulation of Innovation* (Oxford: Oxford University Press, 2006); S. D. Anderman and J. Kallaugher, *Technology Transfer and the New EU Competition Rules – Intellectual Property Licensing after Modernisation* (Oxford: Oxford University Press, 2006); M. Anderson, *Technology Transfer – Law, Practice and Precedents*, 2nd edn (London: Butterworths, 2003).
18. See for instance the home page of the programme's website <http://ec.europa.eu/governance/better_regu-lation/index_ en.htm> and specifically '*A 2nd strategic review of Better Regulation in the European Union*', Commission communication – COM (2008)32 (30 January 2008).
19. A provisional version of this chapter, drafted for advisory purposes, has been offered to the European Commission and is posted by it at http://ec.europa.eu/competition/consultations/2009_horizontal_agreements/index.html (our contribution is entitled 'University of Twente') and is reached directly at http://ec.europa.eu/competition/consultations/2009_horizontal_agreements/utwente_en.pdf.
20. OJ 18 December 2010, L 335/36 respectively L 335/43. The renewed Horizontal Guidelines are also published: Guidelines on the applicability of Article 101 of the Treaty on the Functioning of the European Union to horizontal co-operation agreements, Official Journal C11, 14 January 2011, p. 1 (Corrigenda: C33, 2 February 2011, p. 20).
21. COM (2010) 543 final, de dato 8 October 2010: 'Smart Regulation in the European Union', p. 2.

2
Regulation, Innovation and Competitiveness

Shawn Donnelly

2.1 Introduction

Regulation comprises the rules that apply to private and public actors and the steering mechanisms that are employed to interpret and apply them. The focus is traditionally on the limitation of private activity in the public interest. It constitutes an important component of the business environment in which enterprises work and attempt to improve their business advantages comparative to other firms, including through the development of new and innovative technologies.

That technological innovation is not a goal in itself, but rather the means toward an end (jobs and prosperity), whereby there are several alternative paths to securing that end. Although the path is dictated less than previously by the initial allocation of assets within a country or region, there are limits to the ability of regulation to attract all factors of production in quantity and quality sufficient to guarantee a competitive place in the economy. There is therefore an intense interest in regulation and comparative political economy studies in the institutional foundations of comparative advantage. The choice of regulatory infrastructure, of regulatory rules and of accompanying regulatory cultures favours some forms of economic activity over others while providing for the public interest in various ways. In short, specialisation in regulation goes together with specialisation in economic activity and innovation.

These different paths focus primarily on the choices between incremental innovation and revolutionary innovation, between real and virtual production of goods with value (seen in the contrast between countries whose economies grow as the result of technological development in engineering and those that grow as the result of intellectual property and innovation (ranging from financial services-based on their utility to mobile ring tones and computer games based on their perceived social worth and artistic value)), and a third group that grows through the development and marketisation of virtual products like financial derivatives, whose sole value (tradeability) lies in the perceived utility of the product for a particular end.

A key purpose of this chapter is to show how regulation necessarily makes choices that favour one outcome over others, and how 'smart' regulations can be better understood as constructing rules about which sets of rules apply to a certain situation so as to maximise the economic benefits. Another key purpose is to underline that the choices made here on direction are embedded in publicly validated understandings of the social purpose of public policy and of private enterprise. This may make some kinds of innovation more suitable to country-specific political choices, so that specialisation occurs.

Competitiveness reminds us that the process of achieving economic gains always takes place in an environment in which competitors may potentially poach the source of one's success by copying one's product and selling it at a lower price, by building on trends established elsewhere and translating them into products that are more attractive to consumers, even at a higher price, by blocking the sale of innovative products through private and public protectionist measures, and as well by making more extensive use of intra-industry cooperation between firms to ensure superior technology and products.

This chapter draws on economic theory to examine the means by which regulation can enhance the business opportunities available to private actors. It discusses technological innovation in the context of the contributions that regulation can make to economic development. The five ways in which regulation can foster competitiveness and innovation are by removing regulatory restrictions, reducing transaction costs (through product standardisation and harmonisation of regulatory standards), attracting factors of production, ensuring monopolistic competition, and by preventing the suffocation of innovation by firms with a dominant position in the marketplace. These methods are the subject of section 2.2.

2.2   Regulation and economic competitiveness

Regulation is an issue of intense discussion in public policy due to the effects it has on the business environment. Although business is well known in the current era for maintaining that better or smarter regulation means less regulation, that innovation is best promoted by a freedom of obligation on the part of business to conform to regulatory criteria, there is no direct relationship between minimal regulation and economic success through technology. Indeed, business often benefits from regulation that improves its chances of translating good ideas and products into state-of-the-art technologies that generate robust profits and employment. This can be done in several ways:

- attracting factors of production
- ensuring monopolistic competition
- removing restrictions to market entry
- reducing transaction costs
- preventing technological suffocation

This section discusses these ways in which regulation improves the chance of technological innovation and its translation into economically competitive products. It also discusses the choices involved in attracting the factors of technological production.

Attracting factors of production

All production requires factors of production, which include investment capital, labour (including special skills known as human capital) and other inputs, which may be physical (land and raw materials, pre-fabricated inputs) or intellectual (creative) in nature. Ricardian economic theory underlines that these factors are unevenly distributed, giving economic advantages to some places over others (Bhagwati, Panagariya and Srinivasan, 1998: 29–53). Some of these factor distributions can be attributed to chance, such as the allocation of raw materials, which gives a price advantage, whereas some others can be attributed to the historical gravitation of intellectual, skill and financial resources to particular places, as well as the locking in of those geographical concentrations through the dense network of companies and employment that generate labour and business-to-business markets that are wide and deep enough to attract and hold skilled labour and industry-specific specialist enterprises. The result is the formation of particularly innovative centres, which have a head start on other regions by virtue of the concentration of production factors found there.

The literature on economic competitiveness deals partly with the question about whether and how the distribution of production factors can be changed to benefit regions that are under-developed. Some argue that the act of innovation itself can release regions from the negative consequences of poor asset specificity, which is why technological innovation is often a mainstay of regional development projects for poor regions, but there is also a widespread and long-standing recognition among public policy makers and business that it is equally important to secure access to financial capital and skilled labour, both of which are notoriously mobile and picky.

The investment side is handled first by reducing the regulatory transaction costs to investors that are bringing their capital into an enterprise (including tax rates), and by enhancing the legal certainty for investors that the return on their investment is as secure as possible, which in this case means reducing the risk of politically motivated changes to regulation after an investment is made. These two aspects factor into the political and regulatory risk analysis that an investor conducts before committing to a place.

The labour side is handled by making it as easy as possible for skilled labour to move from where they are to where they are needed (which touches on policies of immigration, residence status and associated benefits of employment, such as tax and pension arrangements), and in some cases where production of a competitive product mandates less the skill of particular individuals and more the above-average technological and production skill of the labour force as a whole, the regulatory establishment of vocational standards and certification. The latter ensures that businesses can rely on a basic standard of skill for those required to carry out sophisticated production techniques.

The capacity for attracting factors of production and channelling them toward innovation and renewal depends as well on motivated political actors (Eisinger, 1989; Yu, 2003) and on the institutional environment that policy makers have to work with (Birch, MacKinnon and Cumbers, 2010).

Ensuring monopolistic competition

One of the iron rules of competition, for a business and not a classical economist, is that too much is bad for you (Chamberlain, 1962). Companies invest money developing new products in the hope that theirs will have something that other products do not, which will give it an inherent competitive advantage. The cost-benefit calculation that a company makes before it invests in new technologies and related production lines therefore includes the questions: is this product so innovative by virtue of the technology and skill involved that it cannot be copied by competitors (Grossman and Helpman, 1991; Vernon, 1966)? How large and elastic is the demand for products in this category (Dixit and Krugman, 1977: 308)? If the product's technology is so advanced that it can protect the company's market share, then the issue of IP rights, to producing a product in precisely such and such a way, with a particular effect, takes care of itself. Eventually this position erodes as competitors successfully imitate. The point at which this is expected to happen is particularly important for regulation questions, since the original innovator must then react to this evolution of the product cycle by moving on to other products, or by relying on regulatory protection of their innovative processes and products through patents (Helpman, 1993; Yang and Maskus, 2001). This is valid both for the producer of an end product, such as a computer, and for the producer of inputs, such as a processor or a graphic card that is built into the end product. Each form of IP protection can be tailored to ensure that the company feels comfortable investing in R&D, and in training to ensure the adequate supply of specialists who can maintain and develop those innovations further. IP rights are particularly important for slowing down the natural development of the product cycle, so that the return on investment can be extended into the future, even though competitors are technically and physically able to reproduce what has been done.

A further approach to innovation and IP rights underlines the endogeneity of being able to recognise and exploit opportunities on the basis of existing knowledge that is protected by IP rights (to prevent predation by investors) (Shane, 2000).

A final form of monopolistic competition occurs through branding, style development and product design, which also depend on the creation and enforcement of company rights to branding, product and process naming in particular. Style development and product design are rarely the subject of regulation, but branding, which refers to the name and its association with certain qualities the product is supposed to possess, is subject to regulation as well, so that only the company which developed a product may use the name or the logo associated with a certain product. Again, this is a commonly used tool that first becomes interesting for a company and a government once it is possible for competing companies to copy the product in exactly the same way and undermine the profits of the original firm or set of firms. It is commonly used for fashion items, ranging from consumer sports goods under popular brands (e.g., Nike or Reebok shoes), to luxury goods (e.g., Luis Vuitton or Rolex products) as well as names for particular regional food products, such as champagne, calvados or parma ham, all of which must be produced in those regions to carry legally the corresponding title, other wine appellations and for non-territorial designations of quality such as *reserva* for wine. Although the former examples are common in the North Atlantic area, Japan and the Antipodes, the latter are specific to the European Union, with its system of regional classification system for regional agricultural products and other products of traditionally distinctly regional origin (Parrott, Wilson and Murdoch, 2002). The protection of branding is not only used for products that are standardised, but also to indicate the qualitative difference that a customer should expect when buying that product.

With the spatial splitting up of research and development on the one hand and production on the other, by which R&D takes place in an post-industrial economy and production takes place in an emerging market economy, regulation also ensures monopolistic competition despite the introduction of variable prices for the same product, known also as perfect pricing, in which companies sell to prosperous consumers at higher prices than they do to less prosperous consumers. This is not achieved individually, but regionally, so that companies forbid the import and sale of the same product from a low-price area to a high-price area. In a time of low transportation and communication costs, which make shopping and delivery easy to do globally, companies then have to rely on regulatory prohibitions on the sale (and import in the course of regular individual travel) of so-called grey-market goods.

Preventing technological suffocation through dominant position

The other side of the regulatory equation is that too little competition is bad for other companies, and by extension, the economy as a whole. When a company develops a dominant position in the market, it can reduce innovation by other companies that wish to develop their products in a different way. This need not occur through coercive action of the company with the dominant position (although this is not uncommon), but the sheer inability of consumers to make real choices limits the size of the market artificially. An example that has been the subject of regulatory action against Microsoft, for example, has been that software developers have found it difficult to make products that are not constructed specifically to work well with the Microsoft operating system Windows. Furthermore, Windows has used its connections with computer manufacturers to not only offer the operating system, but also to offer a host of user applications within the factory-delivered product that directly compete with other products theoretically available to end users. Such practices have been curtailed through regulatory action to prevent technological suffocation through dominant position. An exception to this trend remains in the protection of IP rights (Arezzo, 2006).

Removing restrictions to market entry

The other side of the competition spectrum seeks to break down barriers to trade with regulatory measures. The barriers to market entry may be private in nature or public. Private barriers to market entry include cartel agreements, predatory pricing and access conditions to network industries such as telecommunications, energy grids and operating systems. At the same time, regulation that fosters innovation will also try to avoid rent seeking (Olson, 1984; Kirzner, 1985) and remain as generally pro-entrepreneurial as possible (Sorensen, 2007), an attitude which pervades new public management's focus on clients and customers rather than regulatees.

Reducing transaction costs

Regulation can both impose and reduce transaction costs for business and technological innovation, depending on how they are employed. The most common focus to date is on the cost of regulation in relation to the expected benefits, which is why regulatory impact assessments have made such headway in the UK both as part of the legislative process and as part of the movement in the late 1980s to new public management, which purports to treat regulation as a service, and companies and citizens as customers rather than the more traditional objects of regulatory implementation and enforcement.

Regulation can also be used to promote lower transaction costs for industry through product standardisation and block exemptions for high-tech firms that share R&D and otherwise cooperate for mutual competitive advantage (approved for reasons of national economic competitiveness). The first, standardisation of products and processes, is one that has a weak history due to the resistance that competing companies put up against it. They do this in the pursuit of monopolistic profits that accrue if the product succeeds at winning over the market and being the standard. A recent example is the competition between two different formats for disc-based storage of entertainment media: the now-defunct High Definition DVD (HD DVD), which lost out to the BluRay Disc system. The consequence is that periods of technological innovation are characterised by high degrees of economic uncertainty and 'waste'.

There are strong reasons for regulators not to intervene with the prospect of standardisation in contests such as these (above all that regulators lack sufficient knowledge and are not suitable participants in the creative process), and there are open questions about whether there is any need to promote standardisation through regulations.

Standardisation at this stage has been largely limited to issues of personal and state security, such as parental control systems and so-called V-chips in American hardware that permits state surveillance of personal communications for national security purposes (Balkin, 1996), but also attempts to attain diffuse goals of social control (Johnson and Post, 1996; Goldsmith, 1998). Nevertheless, in those cases where standardisation does occur for product components that are produced widely for a national economy, there has been a strong, positive impact on the economic competitiveness of companies and their products across the board. It makes collaboration across companies easier and engineering knowledge more fungible across companies within a sector, which in turn helps the process of innovation. This has been documented for the automobile industry in Germany, which has been able to keep and build on a common store of knowledge, technology and skill as a result of standardisation (Reich, 1990), but in principle applies to any sector, including high tech.

The modern version of reducing transaction costs through standardisation is to allow ad hoc or even block exemptions from competition policy restrictions to companies that cooperate or collaborate on the development of new products and use them as with inputs in their final products. A good example of this is the US government-sanctioned collaboration between IBM, a computer manufacturer and Intel, a microprocessor manufacturer, on the grounds that national economic competitiveness against Japanese electronics producers was required to hold on to both market share and a technological edge in product development.

2.3   Regulation and the form of technological innovation

Different kinds of regulatory environments promote certain kinds of technological innovation and discourage others, with the result that specialisation and symbiosis between regulation and innovation systems sometimes occurs. Early literature on the varieties of capitalism (Crouch and Streeck, 1997; Hall and Soskice, 2000), expected that this specialisation and symbiosis would take place at the national level (and therefore across countries) due to the national applicability of law and regulation. It divided the world into two groups: liberal market economies and coordinated market economies. Liberal market economies were thought to generate business environments that focused on volatile, short-term revolutionary innovation, change and profit using risky and unproven technologies. Companies, their suppliers, investors, customers and employees were shown to have arms-length relation-ships. This contrasted with the view of coordinated market economies, in which regulation was used to promote and stabilise long-term, incre-mental innovation using, building on and adapting proven technolo-gies. Companies and their stakeholders were thought to have long-term contractual relationships that allowed each of them to specialise over the long term, to engage in long-term planning that generated solid advances rather than extremely risky ones, and generate products of much higher quality, since suppliers and employees all have the security of contract to invest in performing well for a particular company or nar-row economic sector. Even when subjected to the liberalising forces of globalisation or Europeanisation, these varieties are expected by some to persist in important ways (Menz, 2005; 2009). A consequence of this theory of comparative advantage was that countries should specialise in one sort of innovation or the other and avoid mixed systems, which would prove suboptimal for both, and destroy economic innovation in a globally competitive economic environment.

In practice and in theory, there are good reasons to disregard these proposals, even if the effects of regulation on technological and prod-uct competitiveness are true in principle. The varieties of capitalism literature has been criticised for moving too strongly in the direction of parsimony over knowledge of factors that differ by country (Whitley, 1999; Djelic and Quack, 2005), for ignoring a wider variety of capital-isms (Amable, 2003) for looking at a very short period of history during the mid-20th century when these conditions prevailed (Blyth, 2003), for failing to recognise deviation from the pattern generally (Goodin, 2003), and more recently, for failing to deal with the phenomenon in which coordinated market economies in particular have liberalised despite laws and regulations (Streeck and Thelen, 2005). It also fails to recognise the impact of state policies in liberal market economies that reproduce stable planning relationships in strategic, mostly military defence sectors, where high-tech development is key. In cases such as Boeing, stable government contracts provide the starting point for Boeing to invest in human and physical capital necessary to achieve the same kind of progress that would otherwise be present only in a coordinated market economy. What many of the critics could articu-late more clearly, then is that competitiveness ultimately is a firm-level phenomenon for which regulation provides an enabling or disabling environment.[1]

The task for regulators therefore becomes one of diversifying regu-latory instruments to fit these various needs. This starting point underlines the economic argument in favour of diversifying national economies, a point that is central both to portfolio theories of investment, optimal currency area analyses that look at the relative strengths of economies, and economic risk analysis.

Contrary to the varieties of capitalism literature, then, the key question today becomes, therefore: how can regulation be used to capture diverse forms of technological innovation, balancing the portfolio of innovation between unproven and proven technological bases, and in ways that translate fundamental technological innovations (at which liberal market economies have better success rates) into marketable products (at which coordinated market economies excel far better) (Sapir, 2004; Landy, Levin and Shapiro, 2007; Rodrick, 2007). In other words, do we move from national varieties of capitalism to sectoral varieties of capitalism or regional varieties of capitalism, or both (Crouch, 2005; Campbell and Pedersen, 2007)? How can this be done without undermining the regulatory effectiveness of the state across sub-systems?

The most traditional of these variants is regional differentiation, in which different rules apply, so that highly regulated areas are interspersed with lightly regulated ones, as in the case of Hong Kong within China. This effectively reproduces offshore forms of regulatory liberalisation within the borders of a country (Palan, 2003).

It is also possible for a country to liberalise by economic sector, for example, instituting liberalisation for the software industry or coordinated contracts for shipbuilding. It is also possible to liberalise regulations for small and medium-sized enterprises while leaving more intrusive coordinated rules for larger corporations. Within liberal market economies, there is a lot that the entrepreneurial state can do beyond the type of patronage that the American government extends to Boeing and other defence companies. Tax expenditures, public infrastructure and tendering can all be used to smooth the flow of income to companies in particular sectors that are deemed of particular importance. Indeed, as Gregerson (2010) notes,

> regulation is a balancing act in which just the right amount of support and restriction of private sector innovation is required for optimal results.

2.4   Modes of regulation and business innovation

An increasingly important part of the innovation and regulation literature is the consideration given to the mode of regulation and the impact that it has on the degree and type of innovation in the economy. This is apparent in the rise of the Australian school of regulatory theory, and its corresponding adoption in Europe, America and in the OECD as a dominant paradigm, in principle, of how regulatory relationships might best be conceived. There are three main streams of thought in this literature. First, there are those who believe that a regulatory relationship rather than regulatory commands might generate smarter regulation with fewer rules. Second, there are those who point out that a regulatory relationship is better suited to managing the risks of new, experimental and possibly poorly understood technologies without suppressing their development. Finally, there are those who argue from a public choice perspective that state-directed command and control regulation is inherently harmful, due to a number of inherent aspects of regulation that harm public welfare.

At the heart of the Australian school's approach is a rethinking of what regulation is in the first place. Rather than conceived of as a set of state rules and enforcement mechanisms that set out standards and enforce compliance (Hutter, 1997), regulation is perceived as a relationship between regulators (Ayres and Braithwaite, 1992), who act on behalf of the public interest but may not be part of the state, and regulatees, who might not comply with regulatory standards for a variety of reasons. In other words, the Australian school steps back from many of the Weberian assumptions of hierarchical authority and state control that pervade the traditional understanding of the regulatory relationship. Instead, it is multifaceted and with more nuanced understandings of what is best to be done to modify the behaviour of regulatees. The dominant understanding is no longer of separate but of intermingled spheres is now known as regulatory capitalism (Braithwaite, 2008; Levi-Faur, 2005).

The regulatory relationship may rely on a number of mechanisms and for the Australian school it relies on an inclusive spectrum. John Braithwaite's regulatory pyramid has become the standard for conveying not only the tools with which a regulatory relationship may be conducted, but also the optimal mix that should be employed. Command and control measures have their place at the top of the pyramid for purposes of enforcement, but are to be used sparingly, both for the sake of the state, which has limited information and resources at its disposal to act prudently and without causing harm itself, and for the sake of businesses, which then have less to fear from heavy-handed enforcement. This is what Ian Ayres and John Braithwaite name the 'benign big gun' (Ayres and Braithwaite, 1992: 29). It generates best results when combined with different mechanisms of sanctioning, socialising and reinforcement, 'from punitive social control toward moralising social control' (Braithwaite, 1989: 181). This resonates with American writings on the shotgun behind the door, and European writing on the shadow of hierarchy, and thinks through different instruments that might be used instead of shutting down a business. In descending order of intensity, Ayres and Braithwaite name the following possibilities to punish compliance: licence revocation, licence suspension, criminal penalty, civil penalty, warning letter, persuasion (1992: 35).

Beyond this, however, Braithwaite's later work incorporates insights on information gathering on the part of the state and learning on the part of regulatees. Here, the regulator begins with information exchange with the regulated, then goes on to attempt persuasion, followed by various forms of assistance and guidance for the regulated in complying with the spirit or principles of the law, followed in turn by regulators insisting on certain forms of behaviour, and followed up ultimately by various sorts of sanctions (Braithwaite, 2008).

The idea behind this is to ensure that the public interest is guaranteed at the end of the day without strangling off economic activity, particularly innovative activity. In these fast-moving situations, rules-based regulation may have significant welfare costs that threaten economic activity and competitiveness, whereas principles-based regulation allows regulators to apply the spirit of the law without being overly burdensome. In practice, regulators are expected not only to engage in regulatory impact assessments as part of preventing an unnecessary burden on industry, but also in a 'smarter' way, offer guidance and interpretations of the regulatory principles as part of that educating and socialising process within the regulatory relationship. The latter, for example, has proved important in the financial services industries and corporate sectors, where International Financial Reporting Standards are based on principles, supplemented by guidance documents rather than formal, rules-based interpretations by the appropriate body, the International Accounting Standards Board (Donnelly, 2007; 2010).

2.5 Discussion

There are different means by which regulation may assist or suppress economic and technological innovation. Regulation sets the business environment in which innovation takes place. It is therefore in a position to affect the capacity of companies to attract appropriate kinds of investment, including the fixed and human capital to innovate.

It determines the means and costs of market entry or of protection for a company's investment. It determines the capacity of companies to collaborate for technological innovation where this is desired. Finally, it helps determine the extent to which innovation is of the radical, risky variety, or of an incremental, skilled type. This is not only of interest to the division of labour across countries but across companies and regulatory sub-systems as well.

The latter is particularly a point for the comparative advantage of advanced economies, many of which have lost to East Asia the capacity to advance incrementally based on planning relationships between companies, between companies and investors, and between management and employees. The reasons why East Asia rather than Latin America has flourished in this area are multifaceted, but include the strong state-directed economy in many countries of the region (Wade, 2003; Weiss, 1998). Regaining the capacity for this kind of innovation in the West will require attention as to how regulation supports and enables these kinds of planning relationships in key sectors.

Regulation that affects innovation is a balancing act in a number of ways, for which there are no simple answers. How much freedom of manoeuvre should be granted to companies to collaborate or receive protection for their inventions without damaging downstream innovation? How much dissonance in public values can be tolerated in terms of the balance between consumer and producer interests, between manager and employee interests, and between shareholder and other interests in the pursuit of innovation (Donnelly, 2010)? What role should other considerations of the public interest play, if any?

Notes

1. See Dalum, Johnson and Lundvall 2010 on the cumulative, path-dependent nature of knowledge and technology, which requires an institutional environment, education and training – knowledge isn't just radical and accidental.

References

Amable, B., *The Diversity of Modern Capitalism* (Oxford: Oxford University Press, 2003).

Arezzo, E., 'Intellectual property rights at the crossroad between monopolization and abuse of dominant position: American and European approaches compared', *John Marshall Journal of Computer and Information Law*, vol. 24 (2007), pp. 455–xxx.

Ayres, I. and J. Braithwaite, *Responsive Regulation: Transcending the Deregulation Debate* (Oxford: Oxford University Press, 1992).

Bhagwati, J., A. Panagariya and T. N. Srinivasan, *Lectures on International Trade*, 2nd edn (Cambridge: MIT Press, 1998).

Balkin, J. M., 'Media filters, the V-chip and the foundations of broadcast regulation', *Duke Law Journal*, vol. 45 (1996), pp. 1131–75.

Birch, K., D. MacKinnon, and A. Cumbers, 'Old industrial regions in Europe: A comparative assessment of economic performance', *Regional Studies*, vol. 44 (2010) no. 1, pp. 35–53.

Blyth, M. 'Same as it never was: Temporality and typology in the varieties of capitalism', *Comparative European Politics*, vol. 1 (2003), pp. 215–25.

Braithwaite, J. *Crime, Shame and Reintegration* (Cambridge: Cambridge University Press, 1989).

Braithwaite, J. *Regulatory Capitalism: How It Works, Ideas for Making it Work Better* (Cheltenham: Edward Elgar, 2008).

Campbell, J. L. and O. Pedersen, 'Institutional competitiveness in the global economy: Denmark, the United States and the varieties of capitalism', *Regulation and Governance*, vol. 1 (2007), pp. 230–46.

Chamberlain, E., *The Theory of Monopolistic Competition: A Re-orientation of the Theory of Value* (Cambridge, MA: Harvard University Press, 1962).

Crouch, C., *Capitalist Diversity and Change Recombinant Governance and Institutional Entrepreneurs* (Oxford: Oxford University Press, 2005).

Crouch, C. and W. Streeck (eds), *Political Economy of Modern Capitalism* (London: Sage, 1997).

Dalum, B., 'Public policy in the learning society', in *National Systems of Innovation* (London: Anthem Press, 2010), pp. 293–316.

Dixit, A. and P. Krugman, 'Monopolistic competition and optimum product diversity', *American Economic Review*, vol. 67 (1977) no. 3, pp. 297–308.

Djelic, M.-L. and S. Quack, 'Rethinking path dependency: the crooked path of institutional change in post-war Germany', in G. Morgan, R. Whitley and E. Moen (eds), *Changing Capitalisms?* (Oxford: Oxford University Press, 2005), pp. 137–66.

Donnelly, S., *Regimes of European Integration: Constructing Governance of the Single Market* (Oxford: Oxford University Press, 2010).

Donnelly, S., 'The International Accounting Standards Board', *New Political Economy*, vol. 12 (2007), p. 1.

Ebner, A., 'Entrepreneurial state: the Schumpeterian theory of industrial policy in the East Asian "Miracle"', in U. Cantner, J.-L. Gaffard and L. Nesta (eds), *Schumpeterian Perspectives on Innovation, Competition and Growth* (Berlin: Springer, 2009), pp. 369–90.

Eisinger, P. K., *The Rise of the Entrepreneurial State: State and Local Economic Development Policy in the United States* (Madison: University of Wisconsin Press, 1989).

Goldsmith, J. L., 'Against cyberanarchy', *University of Chicago Law Review*, vol. 65 (1998) no. 4, pp. 1199–250.

Goodin, R. E. 'Choose your capitalism?', *Comparative European Politics*, vol. 1 (2003), pp. 203–13.

Gregersen, B. 'The public sector as a pacer in national systems of innovation', in U. Cantnor, J.-L. Gaffard and L. Nesta (eds), *National Systems of Innovation* (London: Anthem Press, 2010), pp. 133–50.

Grossman, G. and E. Helpman, 'Endogenous product cycles', *The Economic Journal*, vol. 101, (1991), pp. 1214–29.

Hall, P. A. and D. Soskice (eds), *Varieties of Capitalism: The Institutional Foundations of Comparative Advantage* (Oxford: Oxford University Press, 2000).

Helpman, E., 'Innovation, imitation, and intellectual property rights', *Econometrica*, vol. 61 (1993), pp. 1247–80.

Hutter, B. M., *Compliance: Regulation and Environment* (London: Sage, 1997).

Johnson, D. R. and D. Post, 'Law and borders – the rise of law in cyberspace', *Stanford Law Review*, (1996), p. 1367.

Kane, E. J., 'Accelerating inflation, technological innovation, and the decreasing effectiveness of banking regulation', *The Journal of Finance*, vol. 36 (1981) no. 2.

Kirzner, I. M., *Discovery and the Capitalist Process* (Chicago: University of Chicago Press, 1985).

Landy, M., M. Levin and M. Shapiro (eds), *Creating Competitive Markets: The Politics of Regulatory Reform* (Washington DC: Brookings Institution Press, 2007).

Levi-Faur, D., 'The global diffusion of regulatory capitalism', *Annals of the American Academy of Political and Social Science*, vol. 598 (2005), pp. 12–32.

Menz, G., *Varieties of Capitalism and Europeanization* (Oxford: Oxford University Press, 2005).

Menz, G., 'Varieties of capitalism and Europeanization: national response strategies revisited', in Gamble, A. and D. Lane (eds), *The European Union and World Politics* (Basingstoke Palgrave Macmillan, 2009).

Olson, M., *The Rise and Decline of Nations: Economic Growth, Stagflation and Social Rigidities* (New Haven: Yale University Press, 1984).

Palan, R., *The Offshore World: Sovereign Markets, Virtual Places and Nomad Millionaires* (Ithaca, NY: Cornell University Press, 2003).

Parrott, N., N. Wilson and J. Murdoch, 'Spatializing quality: regional protection and the alternative geography of food', *European Urban and Regional Studies*, vol. 9 (2002) no. 3, pp. 241–61.

Reich, S., *The Fruits of Fascism: Post-War Prosperity in Historical Perspective* (Ithaca, NY: Cornell University Press, 1990).

Rodrick, D., *One Economics Many Recipes: Globalization, Institutions, and Economic Growth* (Princeton, NJ: University Press, 2007).

Sapir, A (ed.), *An Agenda for Growing Europe: The Sapir Report* (Oxford: Oxford University Press, 2004).

Shane, S., 'Prior Knowledge and the Discovery of Entrepreneurial Opportunities', *Organization Science,* vol. 11 (2000) no. 4, pp. 448–69.

Sorensen, J. B., 'Bureaucracy and entrepreneurship: Workplace effects on entrepreneurial entry', *Administrative Science Quarterly*, vol. 52 (2007) no. 3, pp. 387–412.

Vernon, R., 'International investment and international trade in the product cycle', *Quarterly Journal of Economics*, vol. 80 (1966), pp. 190–207.

Vogel, S. K., *Japan Remodelled: How Government and Industry are Transforming Japanese Capitalism* (Ithaca, NY: Cornell University Press, 2006).

Wade, R., *Governing the Market: Economic Theory and the Role of Government in East Asian Industrialization* (Princeton: Princeton University Press, 2003).

Weiss, L., *The Myth of the Powerless State* (Ithaca, NY: Cornell University Press, 1998).

Whitley, R., *Divergent Capitalisms: The Social Structuring and Change of Business Systems* (Oxford: Oxford University Press, 1999).

Woo-Cummings, M. (ed.), *The Developmental State* (Ithaca, NY: Cornell University Press, 1999).

Yang, G. and K. Maskus, 'Intellectual property rights, licencing and innovation in the endogenous product-cycle model', *Journal of International Economics,* vol. 53 (2001) no. 1, pp. 169–87.

Yu, T. F., 'The entrepreneurial state: The role of government in the economic development of the Asian newly industrializing economies', *Development Policy Review*, vol. 15 (2003) no. 1, pp. 47–64.

3
Legal Design of Smart Rules and Regimes: Regulating Innovation

Michiel A. Heldeweg

> *Getting legislation right is essential if we are to deliver the ambitious objectives for smart, sustainable and inclusive growth, set out by the Europe 2020 Strategy*
> European Commission (2010)

> *There are no easy routes to regulatory improvement.*
> Robert Baldwin (2005)

3.1 Introduction

The concept of 'smart rules and regimes' aims to focus attention on legal instruments that foster technological innovation while providing safeguards against technological risks.

The leading question of this contribution is why smart rules and regimes are relevant to policies *fostering innovation* and how they may be the object of a *legal design methodology*. The main objective of this effort is to emphasise that there is more to fostering innovation than deregulation, and to elucidate possible avenues for further research into legal design of relevant rules and regimes.

Firstly, I will present a (Dutch-based) viewpoint on the 'innovation–regulation' relationship (section 3.2). Next I aim to provide clarification (in section 3.3) on the concept of 'smart rules and regimes', especially in relation to technological innovation. Subsequently, some ideas are presented (in section 3.4) on a possible legal design methodology towards the making of 'smart rules and regimes'. Finally, I will draw some conclusions and make suggestions for further research (in section 3.5).

3.2 The 'Dutch (innovation) paradox'

In my inaugural lecture on smart rules and regimes (Heldeweg, 2010), OECD comments (OECD, 2006) on the 'Dutch Paradox' concerning innovation in the Netherlands underpinned the relevance of adequate regulation for enhancing technological innovation. The findings of the OECD may be summarised by the following quote:

> 'An excellent record in knowledge creation, but a mediocre record in innovation activity' – with 'innovation activity' being defined as, '... successful development and application of new knowledge in a new product and/or process'.

Innovation

The OECD survey prompted a study by the Dutch Scientific Council for Government Policy (WRR)[1] called *Innovation Renewed* (WRR, 2008). The leading question in this WRR report is how government policies may improve Dutch innovation-capacity (WRR, 2008: 20). Innovation is regarded as a concept embracing both a *process* and its *results* in terms of a new functionality or a new way of using an existing functionality, both in the private and the public sector. Thus it is a *'complex system'*, which is not only about creating new knowledge and new technologies, but also about changes in organisation, management and labour, towards faster recognition, diffusion and application of (new) knowledge (WRR, 2008: 18). Innovation is not only about *exploration* (esp. inventions) but also about *exploitation* (esp. valourisation of new knowledge and of new technologies). The latter, however, seems to be problematic, also due to a shortfall in *'entrepreneurship'*.

Failure at and in innovation

This shortfall may be linked to various forms of *market failure* concerning innovation. Major causes of this are: reluctance to initiate innovation with *positive external effects*, uncertainty on *returns on investment*, insufficient or slow *knowledge transfer*, small *profit margins* ('lack of slack'), and insufficient *cooperation* between firms (WRR, 2008: 58, 83).

Next to market failure, there is *'systemic failure'* within the innovation process itself (WRR, 2008: 55). A successful exploitation of new technology requires *institutional* changes within markets, such as new production and distribution chains, (accompanying) technical standards and safety protocols. *Existing* institutions, however, may impede innovation, for instance when a new product fails to pass a safety test (for introduction into the market) because its test criteria are based in existing technology.

Government opportunities

Government may have the potential to rectify market and systemic innovation failure. Technological innovation is a public interest, especially in view of global competition, sustainable development and the improvement of public services, such as energy, transport, health care and infrastructures (WRR, 2008: 29, 58). Taken *intrinsically*, government will focus on securing innovation as such, through securing or improving the 'general rules of the innovation game'[2] rather than on any particular technology or related public service. Taken *extrinsically*, government interventions are geared by the promise of a particular technology to either function as a 'breakthrough' or 'general purpose technology' (e.g. electricity, chemistry, electronics, computers and the Internet),[3] or as a means to improve the quality or efficiency of specific public interests (e.g. energy, public transport, health care). Extrinsic attention may lead government to act as a leading or launching costumer, or as an *initiator*, a partner in Public-Private-Partnerships (PPP), a principal in innovative public procurement, and as legislator.

More failure

With regard to fostering innovation, government should operate on the basis of the 'additionality-principle'. It should intervene only when and where necessary (i.e. when markets and social networks fail at innovation), and must retreat as soon as possible (WRR, 2008: 31–2). Furthermore, involvement of private parties or indeed privatisation may be an important tool in the formulation and execution of public interest innovation policy.

The additionality-principle 'responds' primarily to the dangers of 'government failure' relating to innovation. The main examples of this are: lack of scientific knowledge (causing vulnerability to either misjudgements or to 'regulatory capture'); fragmentation of government innovation policies (resulting from bureaucratic divisions and struggles);[4] over-specificity in its requirements on *deliverables* (in terms of timeliness, technical standards, the requirement of universal access – linked to subsidies or procurement – especially when innovation is still immature); and finally, over-regulation, administrative burdens, and inadequate fiscal and socio-economic policies.

A smart approach?

Clearly (apart from over-regulation) 'dumb regulation' must be avoided; especially technology-dependent regulation (Van Klink and Prins, 2002: 37–8). Such may be the case in rules on market admittance of substances and products, in normalisation within production and distribution chains and networks, in subsidy systems or public procurement strategies, and in regulatory prohibitions which aim to safeguard (especially) the environment, public health and public safety.

Still, avoiding 'dumb regulation' does not cover the full spectrum of legal strategies that 'smart rules and regimes' may provide. A broader and more systematic perspective is necessary. With smart rules and regimes in mind, we may relate to Roger Brownsword's three types of channelling conduct 'x' (Brownsword, 2008: 19–21). He distinguishes negative channelling ('So that agents shall not do x'; presupposing 'a rule that prohibits x'), positive channelling ('So that agents shall do x'; presupposing 'a rule that requires x') and neutral channelling ('So agents may or may not do x' – as they prefer; requiring 'a rule that permits x'). We may substitute 'conduct x' with any particular kind of technological innovation (as applying 'technology x') and similarly channel the regulatory take on this particular kind of innovation as follows:

Table 3.1 Regulatory channelling of innovation

Regulatory channelling of innovation (cfrm. Brownsword)		
Type of channelling	Description of conduct	Typical norm of conduct
Negative	'Shall not do Innovation x'	Prohibition
Neutral	'May (not) do Innovation x'	Permission
Positive	'Shall do Innovation x'	Requirement

Brownsword adds four comments to this threefold 'regulatory range'. Firstly, that a 'negative channelling' (i.e. prohibition) leaves open the decision on modes of enforcement (penal sanctions, tort or administrative redress).[5] Secondly, that 'neutral channelling' (i.e. conduct being permitted) leaves open whether this comes with 'reservations' ('permission with negative reservation') or with encouragement ('permission with facilitation'). Thirdly, that in the absence of a blanket prohibition or unvarnished permission, we are likely to find a regulatory mix of public and private law rules. Fourth and finally, that regulation will hold a 'default position', called regulatory tilt, which projects a normative guideline if regulatory ambiguities need to be resolved or when regulation is silent on a point. In these cases the default position will either indicate prohibition or permission – ambiguity is resolved and silence interpreted assuming the tilt to be either against permission or against prohibition.

Clearly, this approach presents different avenues in which smart rules and regimes may be employed, with a view on fostering technological innovation, and will be of interest to a legal design methodology. This calls for a further analysis of these three options, set out below, in which conduct towards 'x' may be understood as conduct favourable to technological innovation (the norm-object), and 'y' depicts the agent(s) addressed (the norm-subject).

'Prohibitive'. In the prohibitive ('y shall *not* do x') perspective, the focus is on risk regulation: 'regulating technological risk' (Fisher, 2010: 6, 7), that is, the regulatory safeguarding of public interests in a clean environment, public health and public safety, against (outside) threats – prohibitions, but also 'permissions with *negative reservations*'. This focus, however, must come with a view on how, while curbing or channelling exploration and exploitation of technology, risk regulation can meanwhile accommodate technological innovation.

This approach we name: 'innovative risk regulation'. In fostering innovation, the choice and design of regulatory restrictions should reflect innovation-efficiency and innovation-effectiveness.

Innovation-efficiency is about avoiding over-inclusiveness of norms: prohibiting more than is necessary to protect vulnerable interests. Targetedness of the regulated conduct is necessary to avoid such over-inclusiveness (i.e. inefficiency). Detailed regulation may be the answer, but often comes with technical specificities that make restrictions vulnerable to technological innovation the aforementioned problem of technology-dependency.

Innovation-effectiveness is about promoting more desirable innovations by placing restrictions on existing or emerging technologies. Such restrictions could operate as a push towards a technological shift to a new or improved technological functionality. A mere prohibition (of certain technologies) may (similar to a command – e.g. of emission targets) promote innovation, if industry has sufficient (financial) resources and time to make necessary alternative investments, and if a level (competitive) playing field is safeguarded. A '(smart) regulatory mix', for example, may add an encouraging *pull* (the proverbial 'carrot') to the compelling *push* (the proverbial 'stick') such as by combining restrictions with subsidies or tax exemptions for first movers. In effect this shifts the regulatory response from 'negative' towards 'neutral' channelling.

'Permissive'. In a permissive ('y may or may not do x') perspective, our focus lies with creating favourable regulatory conditions towards enhancing technological innovation as a matter of private choice. We name this: 'innovation-facilitating regulation'.

'Permission with positive facilitation' would clearly fall into this category, but also, for example, creating a zone in which private actors 'may' and 'can' indeed (legally) pursue certain innovations, more restricted elsewhere.

The baseline of facilitation is to ensure the presence of a legal infrastructure favourable to technology innovation; based on public interest considerations but geared to foster private interests, so that private actors may act upon expected private 'gains'. A robust and attractive system of property rights (including intellectual property), sound and versatile contract and competition law, as well as trustworthy regimes for recognition of specialised knowledge, skills and performance (e.g. diplomas and certification) are essential.

In addition and more conducive (as 'pull'), some specific legal arrangements, innovation enticing, could provide interesting stepping-stones towards innovation, such as: tradable public rights (e.g. emission allowances), legal grants (e.g. subsidies, favourable loans and securities), (re)allocation of property rights (e.g. removing 'anti-commons' barriers – Heller, 1998), and regulatory competition (Tiebout, 1956).

'Compelling'. In a compelling ('y shall do x') perspective, innovation is a public interest that warrants active persuasion, through regulation that obliges innovative conduct. We name this: 'innovation-compelling regulation'.

Following four types of regulation,[6] this compelling regulatory perspective presents itself in many different examples. Firstly, through direct regulation, where conduct follows 'command and control' and breaches are punishable under law, which features prohibitions and commands aimed at shifting from existing undesirable technologies to new and promising technologies. Secondly, through indirect regulation, where conduct follows efficiency considerations and is sanctioned by – comparative – economic disadvantage, such as by taxation (or by tort law).[7] Thirdly, through self-regulation, where conduct follows social considerations and is sanctioned by criticism and ostracism. Examples of compelling self-regulation generally require that self-regulation transforms into intermediary regulation, in which regulation results from cooperation, such as where government (or a third party) operates as initiator or matchmaker. Public and/or private initiatives towards certification of products, services, processes and organisations (as a prerequisite to a go-ahead), are important examples. Operating as a launching (or leading) customer will require proper schemes of innovative procurement or schemes towards public-private partnerships. Furthermore, *'quangos'* (i.e. independent regulators) or social enterprises (e.g. housing corporations, health care providers and universities) may be the intermediaries that initiate innovative projects (on the basis of a regulatory framework provided by government). Fourthly, the use of inherent regulation, where conduct follows inbuilt or systemic constraints of functionalities,[8] sanctioned by absence or loss of (physical) functionality. These may present themselves – compellingly – through the need to apply (or keep up with changes in) state of the art technology to (optimally) make use of a functionality. A low-tech example is the 'speed ramp', while a high-tech example is adherence to internet protocols in providing and using ICT-services.

Balancing act

Clearly, this regulatory range indicates the balancing act that is needed. Not only must technological innovation be fostered, but there is also the public interest of securing against technological risks. Ideally both interests must be acted upon jointly, indifferent to whether the initiative to act lies with fostering innovation or curbing risks.

Thus, there is an abundance of reasons to look more closely at what can be done to design regulation that can cope with the challenges, which the aforementioned balancing act presents us with. Consequently, focusing on the possibilities of designing 'smart rules and regimes' seems a sensible next step.

3.3   The concept of 'smart rules and regimes'

Conceptualising 'smart rules and regimes' focuses attention on legal forms of regulation and the general *normative context* that underpins and constraints their validity and bindingness. First, the concepts of 'rules' and of 'regimes' must be defined, preferably in simple and yet robust terms.

Rules are defined as linguistic statements, projecting a mode of conduct or a power conferred, which is prescribed or attributed (bindingly)[9] in instances as defined within the rule (or its underpinning rules). In general, rules consist of: a 'subject', that is, the person(s) prescriptively addressed – subject 'y' in: 'y shall/may or can (not) do x'; an *'object'*, that is, the conduct prescribed – conduct 'x' in: 'y shall/may or can (not) do x'; an *operative mode*, that is, the prescriptive modality – prescribing 'shall/may' or 'can' in: 'y shall/may or can (not) do x'; a *'norm-condition'* ('z' in: 'y shall/may or can (not) do x, if and when z') (Ruiter, 2010).

A regime is defined as a system of (such) rules, which in conjunction includes not only norms, but also the mechanisms of decision making and the network of involved actors.[10] A regime holds at least the minimum of objective legal norms necessary to underpin subjective legal relations, such as in a permit or a subsidy scheme, and in (general) contract and property law. Some regimes are abstract (i.e. of general applicability – e.g. contract and property law), and some concrete (i.e. related to a specific interest – e.g. an environmental permit system or indeed an environmental law code).

Wicked problems, smart response

I name rules and regimes smart, in as much as they are more successful in adequately addressing the need for legal regulation in complex circumstances of:

- high *societal or technological dynamics*, which present a desire for adaptable (or self-adaptive) (systems of) legal norms which dynamically enshrine new/improved knowledge (technological innovation) or new/changed opinions/values (social innovation – often following technological innovation);
- major *contrasts or rather conflicts of interests*, presenting (more) fundamental differences in (opinions following underlying) values, which present a desire for (systems of) legal norms, which may mediate tensions between these values, such as between public and private interests, or between opportunity and risk.

A combination of these two circumstances – high dynamics and major conflict – presents a 'wicked (regulatory) problem' (comp. Rittel and Webber, 1973), such as is manifest in debates on technological innovation and the likelihood, magnitude and distribution of their risks and benefits. Clearly, these problems pose a challenge to regulatory governance,[11] that is, to the design of smart rules and regimes, aimed at balancing both dynamics and conflicts.

As to definitions, the concept of smart rules and regimes is not coined to delineate sharply between, smartness and dumbness, but primarily to express the ambition to overcome wicked regulatory challenges.

Smart rules and regimes are clearly relevant in two exemplary cases. Firstly, when fostering services of societal interest, there is a need to balance public and private interests in these services, in the context of

Table 3.2   Dynamics (effective and efficient)

Dynamics (effective and efficient) →			societal	technological	
Conflict ↓(legitimate & legal validity)		–	+	+	–
Private ⇔ Public	–	–/–	–/+	–/+	–/–
	+	+/–	Smart rules & regimes		+/–
Risk ⇔ Opportunity	+	+/–	+/+		+/–
	–	–/–	–/+	–/+	–/–

A '+' or '–' points at presence or absence of a conflict or of dynamics.

dynamic changes in (primarily) political opinion (and/or in commercial strategies), on how and which services need be rendered. Secondly, when fostering technological innovation of societal interest, especially balancing advantages and risks of innovation, in the context of a context of dynamic changes in (primarily) technology (and society).[12] Table 3.2 shows how smart rules and regimes may thus be positioned – the references to legitimacy, legal validity, effectiveness and efficiency will be clarified below).

In this contribution the focus is on fostering *technological innovation*, with a view on (a) strong dynamics (i.e. a stress on robustness and on adaptability) and on (b) avoiding unacceptable risks. Clearly, this touches upon the challenge addressed by, among others, Somsen (Somsen, 2009: 21 – translated by MAH):

> The continuous safeguarding of the topicality of a regulatory regime, measured by the current state of technology, is the greatest and seemingly most hopeless challenge of technology regulation. [...] Effective regulation presupposes a procedural and institutional facility which facilitates a simple and fast adaptability of the regulatory regime to new technological factuality.

In this contribution, smart rules and regimes are discussed in terms of their ability to provide such an 'institutional facility'.

Variables and maxims

The smartness of rules and regimes relates to balancing two variables: high dynamics and strong conflicts of interests. This reflects underlying legal governance dimensions and accompanying principles or maxims of regulatory effectiveness and efficiency (under dynamic pressure) and regulatory legitimacy and legal validity (under pressure of conflicting interests). Clearly, the challenge, as described by Somsen, illustrates this. If legal norms do not 'align' with the current state of technology, they stand to become either (obsolete, hence) ineffective (i.e. inadequate in protecting against risks) and/or (obstacles, hence) inefficient (i.e. unnecessarily hampering the deployment of new technologies). Making these norms more adaptive to technological change may, however, put stress upon their legitimacy (e.g. through involving private rule making) or their legal validity (e.g. through discretionary overstretch in executive use or court interpretation).

Consequently, the design of smart rules and regimes must rest upon guidelines, which do justice to these legal governance dimensions and accompanying principles of good legal governance. These principles reflect both the requirements of the rule of law (as legal norms adherence – safeguarding freedom and fairness between government and citizens and among citizens reciprocally) and of serving the public interest (as governance optimality – ensuring that public interests be served properly and balanced). The latter ('proper public service') should lead to the design of effective and efficient regulatory strategies; the former ('fairness through the rule of law') to the design of legitimate and legally valid rules and regimes.

In this day and age, the term 'regulatory governance' points at the (vertical, horizontal and perhaps diagonal)[13] diffusion of regulatory power, following the postulate of *legal pluralism*, instead of the classic *liberal legal* underpinning (Scott, 2010). Within this context the regulatory governance debate especially addresses: 'the tensions between effective and instrumental regulatory governance, on the one hand, and demands for accountability and respect for process and rights within constitutionalism on the other.' (Ibid.).

Four dimensions, four principles

In keeping with this perspective, I proposed (Heldeweg, 2010) that the smartness of rules and regimes comprises a cumulative judgment on the aforementioned four counts, which I will now clarify.

Legitimacy is about *'which* particular regulatory interventions *can legally* be brought about by *who?'* Its focus is on the *id quod* dimension (as opposed to *modus quo* – in legal validity) of the regulatory power to (unilaterally) bind others. Governmental legal acts are a major example. In the public law appraisal of state and government powers, legitimacy is the prime issue, as it delineates the demarcation between state power and citizens freedom(s). The issue, however, stretches beyond the context of government hierarchy (featuring binding government transactions). Other relevant archetypical 'institutional environments' of regulatory activity, are markets (featuring competition-based transactions) and social networks (featuring collaborative transactions – public or private).[14] Each environment brings its own underpinning of legitimacy, pointedly described in terms of 'voice' (democratic legitimation of government), 'exit' (demand shifting to another supplier) and 'loyalty' (reciprocal willingness to cooperate).[15] As serving public interests may be left to (or be strongly influenced by) private actors, alternative institutional environments with their particular demands on legitimacy will likewise be relevant to the smartness of rules and regimes that shape and operate within these environments and the alignment of structures of governance and accompanying transactions (Van Genugten, 2008; Scott, 2010).

Legitimacy should be understood primarily in legal terms, as the legal power or competence to bind others legally or de facto, especially unilaterally. Even for government this will stretch beyond the concept of legality; in general 'legal authority under the rule of law' seems more fitting.[16]

In the context of technological innovation 'legitimacy' is the dimension determining which agents can authoritatively decide on the introduction and use of new technologies – either by governments (by democratic mandate), by markets (by consumers preferences), by social networks (by consent) or by some hybrid mix of these (all of the aforementioned?), either in public or private law legal forms or concepts.

Legal validity is about '*how* a particular regulatory intervention *may/ shall* or *can legally* be brought about?' It focuses on the *modus quo* dimension of regulation in terms of the availability (and proper use) of regulatory legal tools (e.g. legislation, contracts and permits), respect for higher (written or unwritten) legal norms, and applicability of legal controls (i.e. enforceability and legal protection).

The full merit of legal validity' is realised only if and when we not only consider lawfulness in terms of (likeliness of) passing the test of judicial review by a court of law,[17] but also, as a matter of (publicly debatable) '*justice*', as in congruency with leading legal concepts and principles, such as 'distributive justice' and 'openness' or transparency' (apart from possibilities for a judicial test).

In the context of technological innovation this is the dimension where, in particular, the distribution of risks and benefits of innovation will be tested.

Effectiveness is about '*what* particular regulatory intervention *can practically* be brought about?' It focuses on whether a rule or regime can result in intended (changes of) conduct. Effectiveness is served when rules and regimes adequately fulfil their functional potential to bring about the kind of conduct and/or activities, which is or are deemed beneficial to the particular public interest. With regard to individual norms, the following elements are of (major) importance in ensuring that a rule/regime adequately depicts and prescribes the desired kind of (pattern of) behaviour:

1. choosing the proper 'norm operator' – projecting the normative mode in terms of shall/may (not) or can, fitting to the desired conduct or power;
2. choosing the proper 'norm object' – projecting the conduct or empowerment as a *proxy* that is most in keeping with the relevant public interest (certainly not under-inclusive and, if possible, conducive);[18,19]
3. choosing the proper 'norm conditionality' – determining whether or not the norm is of a categorical or a hypothetical normative mode;[20]
4. choosing the proper 'norm subject(s)' – that is, addressing the proper agents (individual or general). Not only should a purposive fit of norms to public interests be ensured, but (unintended and) undesirable side effects (to the intended or adjacent public interests) must be avoided.[21]

Furthermore, fostering effectiveness requires ensuring, on a regime or system level, 'external coherence' or systemic effectiveness; avoiding a clash with other norms resulting in a conflicting prescription of conduct (conflicting requirements); ensuring certainty (measurability of adherence) and credibility of enforcement (and legal protection).

Relating specifically to technological innovation, dynamic appropriateness must be ensured, safeguarding both robustness through adaptability (of content, interpretation or application) thus avoiding under-inclusiveness. The latter is especially relevant to the effectiveness of rules and regimes in the context of technological innovation and, in a way, is similar to a smart bomb's capability to adjust to a moving target so as to hit a target at least as good as when the target were motionless.

Efficiency is about '*how* a particular regulatory intervention *can* practically be brought about?' It is concerned, firstly, with whether regulating is a 'cost-effective' policy effort to create presumably effective rules or regimes (as policy output). Secondly, it addresses the 'cost-effectiveness' of the rule or regime (as resulting from the regulatory effort) as regards the impact on the regulated conduct (as policy outcomes).

The output efficiency hinges on the relative regulator's costs concerning creation and management of regulation (including informing, enforcing and protecting). The regulator should aim to choose (types of) rules and regimes which (by comparison) incur the least regulatory management *costs* to the regulator (e.g. general rules; empowerment of norm-addressees to enforce among themselves; building upon existing rules and regimes).

The outcome efficiency depends on the relative behaviour adjustment costs of the regulatees (i.e. norm-addressees); for example, administrative burden, (profit) losses – some temporary, some structural. The regulator should choose those (types of) rules and regimes, which cause (by comparison) the least regulatory or administrative burden on regulatees, by avoiding unnecessary 'juridification', over-regulation, over-detailed regulation, (over)fragmentation of (property) rights and powers (i.e. 'anti-commons'), and facilitation of self-regulation. On the level of individual norms these aspects are 'captured' by avoiding over-inclusiveness through a proper choice of 'norm object' (not unnecessarily restrictive), of 'norm conditionality' (if possible hypothetical), 'norm subject(s)' (no more norm-addressees than necessary) – all of which amount especially to targetedness or precision: in 'smart bomb' terms, avoiding 'collateral damage'.[22] On the level of regimes especially 'external coherence' (avoidance of conflicting requirements; fostering complementarities and synergy), but also legal certainty and credibility of enforcement (and legal protection) need be ensured.

Efficiency is here expressed in terms of the least in either regulators' regulatory management costs or regulatees' behaviour adjustment costs – least in a comparison of alternative rules and regimes being equal either as purported norms (effectively expressing the desired conduct – the regulatory output), or as conduct invoked (effectively in conformity with the norms). Efficiency is regarded as a relative (or comparative) concept, and is about whether the choice of a specific rule or regime comes without unnecessary cost or by comparison with the lowest (necessary) costs. 'Smoothness' of management or of behaviour adjustment – 'regulatory ergonomics' – comes with minimising the burden inflicted upon regulators and/or regulatees, and maximises output- and outcome-efficiency. Note though, that lowering costs as a matter of efficiency may (be) extend(ed) to creating or even maximising private benefits in alignment with relevant public interests; the rules and regimes become more efficient (i.e. more easily manageable or adjustable to), but also (and otherwise relevant to both regulators and regulatees) (more) profitable, which supports effectiveness.[23]

From the standpoint of technology innovation '*dynamic* appropriateness' needs to be ensured, safeguarding targetedness through adaptability, focused on substantively safeguarding and, where possible, 'freedom from regulation' or a 'permissive tilt', again avoiding over-inclusiveness. This is similar to how a smart traffic light's adjusts to the absence of crossing cars, so as to avoid unnecessary waiting when there is no risk of crashing into another car.

In the above, the term 'smoothness' was used to describe the efficiency maxim in terms of (producing manageable, but especially of) creating user-friendly norms (the users being regulatees or norm-addressees). The term 'ergonomics' was coined as possibly more fitting, as it expresses the notion of choosing and designing rules and regimes which optimally align with (existing) 'natural' behaviour or governance modes (including relevant (types of) values, preferences, incentives and inter-/transaction patterns).[24] Thus, regulatory ergonomics (or ergonomics of regulatory governance) not only call for a lowering of the regulatory burden, but also the efficient allocation of obligations and (correlative) property rights (or entitlements). This means that the regulatory burden should preferably not be a serious impediment upon desired conduct, but insofar as it is, it should preferably be a burden only upon norm-addressees for whom it is, by comparison, easiest to shift the burden onto others; that is, transfer onto others, preferably as a result of a voluntary transaction (Ruiter, 2008). Furthermore, regulatory ergonomics are also promoted by clarity and simplicity of (systems of) norms, and a responsive/participative style of regulation enhances opportunities for rules and regimes to match norm-addressees' 'particularities'.[25]

Again, the 'how' of the efficiency dimension is concerned with measuring costs or burdens, but especially with comparing alternatives. Clearly, greater efficiency may indirectly enhance the effectiveness of the rules and regimes involved, as well as benefit legitimacy (if only by being less burdensome) and legal validity (since inefficiency may be a sign of disproportionality).[26]

In analysing effectiveness and efficiency we find that some issues are, sometimes 'symmetrically', relevant to both. The first examples of this are 'precision and targetedness', and 'over-regulation', which relate to either over- or under-inclusiveness, amounting to either not reaching goals, otherwise harming the featured public interest or harming other public interests, or being over restrictive, thus unnecessarily limiting freedom and creating an unnecessary regulatory burden on both regulators and norm-addressees.

Another example are the abovementioned regulatory ergonomics, which are relevant in terms of enhancing functionality, especially through enabling (make effective' – such as through tradable public rights) and avoiding transaction costs ('keep efficient' – such as through avoiding fragmentation and non-alignment with behavioural and governance patterns).

A final example of dual relevancy is the issue of robustness, either geared to retaining effectiveness (i.e. adaptability) or efficiency (i.e. targetedness).

Name of the game

Interdependencies between legitimacy, legal validity, effectiveness and efficiency, place the ideal of smartness of rules and regimes on a range between *satisficing* and optimising. As *satisficing* requires meeting minimum criteria for each maxim, not allowing any trade-offs; optimising, although initially geared towards best performance on each count, does also allow for balancing between dimensions or maxims (with respect of minimum criteria). Such balancing will not only need to reflect the nature of a given public interest and the kind of dynamism and or conflict with other interests, but must also relate to the nature of the (existing or proposed) 'institutional environment', as described in the above, of which hierarchy (government), competition (markets) and collaboration (social networks) are the archetypical forms.[27]

In practice hybridity between these forms, as in the example of a commercial enterprise (also) being involved in issuing public law certifications for technologically complex industrial installations, is ubiquitous. Hence, the evaluation of the smartness of the rules and regimes involved will hinge on whether there is sufficient alignment between governance and regulation, in terms of all four indicators. Legitimacy, for example, may suffer considerably when a commercial enterprise is bestowed with public powers, which in effect (e.g. through certification) push out competition – so private interests drive out public interests, or hierarchical powers distort the workings of the market.

3.4   The concept of a legal design methodology (towards 'smart rules and regimes')

Assuming relevance of smart rules and regimes in the face of the fostering of technological innovation, we need to consider their legal design.

A methodology of legal design?

By analogy, examples from industry, fashion and furniture suggest that to create 'legal objects' on the basis of a design (based in a design methodology) makes good sense. According to Ruiter,[28] successful examples of such designs 'designate the projection of a *type* of *artefacts* with a *function* determining their *form*'.[29] Similarly we may possibly create legal artefacts, depicted in their design as a specific *form-function* relationship, capable of performing or fulfilling legal functions.[30] Especially *rights* (transferred as if they are objects), *obligations* (imposed as if they are burdens) and *powers* (conferred as if they are tools) are likely candidates of such legal artefacts – as are property, servitude and inheritance (Ruiter, 2010a: 2).

To arrive truly at a general conceptual framework as a basis for a methodology of legal design, which (general guidelines) could operate in different areas of law (as smart rules and regimes may 'operate' as regulatory mixtures across different areas), we should leave behind the legal dogmatics of specific areas of law, such as civil or administrative law. A promising, more abstract, approach is to use concepts of 'legal theory', more specifically the theory on 'legal institutions' (Ruiter, 1993; 2001), such as marriage, property rights, corporations and public authorities (Ruiter, 2010a: 3–4).

An institutional approach

Legal institutions depict a normative mode of human behaviour in the form of (1) a system of rules (2) projecting a state of affairs (3) that ought to be realised (4) by a social practice regulated by those rules and (5) expressive of a common belief that the state of affairs is (6) actually the case. Element 2 (projecting a state of affairs) readily leads to a classification of different legal institutions, projecting: (1) a legal quality; (2) a legal status; (3) a personal legal relation; (4) an objective legal relation; (5) a legal configuration – and furthermore, (6) (through 'personification') legal subjects and (7) (through 'reification') legal objects. These basic legal forms, delineated as possible legal design objects, may be geared towards providing design guidelines, derived from (1) the general features of all legal institutions and (2) the specific features of legal institutions of different classes (Ruiter, 2010a: 7–10 – also including examples of such guidelines).

Clearly, such guidelines primarily relate to – by analogy – the 'drawing of the dress' (i.e. the design of a legal artefact), rather than the 'dress made' (i.e. a specific instance of such legal artefact). Legal design is, after all, primarily concerned with conceiving, as concepts, possible institutional practices and their corresponding 'systems of legal norms suited to elicit and control such practices' (Ruiter, 2010b: 4). As such the design of concepts of legal artefacts provides the foundation for (guideline-based) conceptualisations of individual instances of such artefacts, suited to a particular situation.

An internal perspective

A methodology of legal (institutional) design should primarily consider the internal structure of 'legal institutions as systems of norms'. A major focus would be to determine how the 'constitutive elements' of such legal systems could properly be framed: the legal norms themselves (Ruiter, 2010c: 1). A fitting methodology could firstly address rules projecting legal norms of conduct and, secondly, rules projecting power-conferring norms legal acts.

As to the design of legal norms of conduct ('x shall/may (not) do y (when Z)'), Ruiter proposes that their design should adhere to the following three guidelines:[31]

1. take into consideration the four basic components of every norm: Subject ('x'), Object ('y'), Operator ('shall or may (not)'), and Conditions ('z');[32]
2. decide to which extent the logical oppositions (in variations of 'shall or may (not) do': command, prohibition, permission and dispensation) are relevant determinants of the design;
3. establish whether the design is such as to generate legal relations (according to Hohfeld) of a 'claim-duty' or 'privilege-no-claim' nature (discriminating between different kinds of rights – 'claims' and 'privileges' – and their legal opposites – 'duties' and 'no-claims'), and, if so, whether the respective rights holder positions and rights addressee positions are properly specified (in terms of unital and multital relations – MAH).

These three guidelines summarise an analysis, which specifies norm-components and presents categorisations of norms, types of rights, and rights holders and addressees.

Framing the norm-components strikes a chord with the above analyses of the dimensions of effectiveness and efficiency, especially in terms of defining 'inclusiveness' (and avoidance of over- and under-inclusiveness) – which makes 'guideline 1.' especially relevant to smart rules and regimes.

Combining the positions of the components object and operator, yields four basic types of norms of conduct ('command', 'prohibition', 'permission' and 'dispensation') together with their reciprocal relations: contradictory, contrary, subaltern or subcontrary. This basic understanding is indispensable, for instance, for the concept of a regulatory range (channelling conduct) and regulatory tilt, as well as for reasons of legal system-consistence (especially relating to effectiveness and efficiency, and indirectly to legitimacy and legal validity) – hence we also need to take 'guideline 2.' into consideration. This is particularly challenging as legal language brings with it the ability to express the 'shall-operator' in terms of the 'may-operator' and vice versa. Thus, a command (an 'ordered act') is the negation of dispensation (the 'allowance not to act'); prohibition (an 'order not to act') is the negation of permission (an 'allowance to act'); permission (*supra*) is the negation of prohibition (*supra*); and dispensation (*supra*) is the negation of command (*supra*). Careful consideration is clearly of vital importance and must come with reflection on systemic aims (such as regulatory tilt) and regime consistency.

Similarly, we may use Wesley N. Hohfeld's four fundamental types of legal rights ('claim', 'privilege', 'power', 'immunity'), within four types of legal relations, as two groups of two types of rights-relations, either as relations based in norms of conduct or as relations based in norms of competence (to change norms of conduct). As to norms of conduct a 'claim' has a 'duty' as its legal correlative and a 'privilege' has a 'no-claim' position as its correlative. The positions of rights holders and rights addressees may be 'unital' (one person of closed group) or 'multital' (an open class of persons), which leads to eight different types of (claim – duty and privilege – no-claim) legal relations. Clearly, considering earlier remarks on effectiveness and efficiency relation to norm-subjects, adherence to 'guideline 3.' is called for.

Next to norms of conduct we must consider power-conferring norms, or norms of competence, which enable one or more agents to create[33] or abolish legal norms (Ibid. 2000c:1). Such a 'norm of competence' requires that we introduce, in addition to the norm-operators 'shall' and 'may', a third norm-operator: 'can', signifying the ability to bring about legal effects especially through the performance of a legal act (e.g. legislation, contracting). Empowerment to perform such legal acts must come from within the legal order and is met, within that order, by the correlative 'liability' of others to adhere.

Similar to norms of conduct (claim and privilege), there is a square of rights and (two) legal relations following from norms of competence: power, vis-à-vis its correlative liability, and immunity, vis-à-vis its correlative no-power). Again, it is relevant to distinguish between (in all eight types of relations based on) unital and multital rights holder and rights addressee positions.

Reflection on (1) norm-components, (2) logical oppositions, and (3) types of legal relations is vital to the making of (regimes of) legal rules, and the specific nature of norms of competence should be well understood. The power to create or abolish legal norms is, of course, vital in order to ensure that norms are adaptable in the face of technological innovation – apart from other mechanisms, such as changing modes of interpretation of abstract terms and conditions. Different classes of competence norms may be categorised, underpinning different classes of legal acts (Ruiter, 2010d: 9–10). Design guidelines should accommodate the making of such different types of norms.

3.5   Conclusions and the way forward

The leading proposition to this contribution has been that smart rules and regimes provide a useful framework towards understanding the regulation-innovation relationship, as well as towards the legal design of such rules and regimes.

The ideas behind such design, especially as phrased by Ruiter, open the perspective of framing smart rules and regimes as legal institutions. Clearly, this translates quite readily into possible guidelines concerning the internal structure of legal institutions, especially norms of conduct, norms of competence and legal acts. Such guidelines may well prove indispensable to addressing the question of how some internal structures may be smarter than others in terms of the four dimensions (and corresponding maxims) as coined in the second paragraph.

In particular, matters of effectiveness and efficiency (dimensions) may be addressed on the design level of legal norms by components such as the determination of the norm-operator, norm-object, norm-subject and norm-condition. Apart from legal consistency, however, an institutional perspective on legitimacy and legal validity is less apparent. As much as effectiveness and efficiency are relevant to regulating technological innovation, so are legitimacy (as in who decides on setting certain standards for new technologies) and legal validity (as in safeguarding legal certainty as a precondition to investing in innovation). Consequently, in terms of legal design, we need a framework or methodology that also addresses these two dimensions and accompanying principles.

Internal and external

At this stage of analysis, we must acknowledge that the focus on the internal structure of legal institutions must be matched by a broader design scope, encompassing both rules and regimes. Legal institutions will often be framed as regimes, forming a conjunction of rules as a system, projecting the institution. The internal make-up of such regimes – how to frame a conjunction of rules – is a primordial challenge to the design of smart regimes.[34] Subsequently, the focus on the internal structure needs to be matched by an external view on the legal consequences of regulatory governance or policy considerations. This seems especially relevant as regards legitimacy and legal validity.

The need for such an external perspective, however, is important to all dimensions and principles. The regulatory governance considerations and proposed strategies which go under the labels of 'Better Regulation' (OECD, 1995 and 2005; EC, 2002, 2005) and 'Smart Regulation' (Gunningham et al., 1997 and 1998; EC, 2010), seem proper 'illustrations' of the impetus, scope and focus of such strategies towards an improved regulatory contribution to (among others, sustainable) innovation and growth. Especially the references by Gunningham et al. to 'regulatory design processes', 'regulatory principles' and 'instrument combinations', hold a promise on these counts – even though their proposals seem more focused on framing the policy process than providing guidelines for legal design.

Design of regulatory channelling and tilt

On a more specific note, the institutional take on legal design and smart rules & regimes may already provide a relevant analytical tool to elaborate methodically on the concept presented in the first paragraph and based on Brownsword's notion of a 'regulatory range' or of 'regulatory channelling'. Applied to (especially fostering) technological innovation, we should, as a matter of legal design, be able to express such regulatory strategies (including reservations, facilitation and tilt) in terms of the squares of normative oppositions (command, prohibition, permission and dispensation), and the contrary, contradictory, subcontrary and subaltern relations between rules fitting to these strategies. Thus a methodologically more sophisticated perspective seems feasible on the possibilities of design rules and regimes that express regulatory strategies becoming technological innovations.

Summing up

So, our quest continues. It will address both the general scope for a legal design methodology and the specific angle of regulating innovation. As to the latter, focal points of a design methodology of smart rules and regimes should be, on the one hand, determining design guidelines with respect to adaptability and ergonomics of rules and regimes, and on the other hand, guidelines concerning mixing of legal instruments as the challenge of building optimal legal regimes, as regards technological innovation and in keeping with maxims of legitimacy, legal validity, effectiveness and efficiency.

Notes

1. Wetenschappelijke Raad voor het Regeringsbeleid. Available at http://www.wrr.nl/english/. Accessed on 25 September 2011.
2. Especially basic requirements (e.g. legal certainty and trust through a basic legal infrastructure) and avoidance of institutional barriers (e.g. technical standards).
3. Expected to trigger, as did the examples mentioned, many other innovations and thus contribute significantly to (economic) welfare.
4. This fragmentation links with the economic concept of anti-commons: a fragmented pattern of many holders of rights in (scarce) resources (i.e. government powers) each of which is potentially prohibitive for pushing ahead with an innovative project, thus causing underuse of the resource (Heller, 1998).
5. Brownsword (2008: 15) also refers to the work by Murray and Scott (2002) in which three dimensions of regulation are distinguished: goal/rule/norm; monitoring; behaviour modification (Murray and Scott, 2002: 500–5).
6. We use the distinction in regulatory modes as presented by Murray and Scott (2002: 500–5) – following especially earlier work by Lessig.
7. Subsidies, loans, securities and tradable allowances, also fall within this group, but these – generally – belong to 'facilitative regulation', whereas here the focus lies on more compelling instruments, such as taxes. In practice, of course, all these instruments may be either more facilitative or more compelling, depending upon specific circumstances and individual preferences.
8. Also referred to as 'design built' or 'architectural' control or regulation (Murray and Scott, 2002), for example, artefacts, machines, products, services, production processes and infrastructures.
9. We take the view here that legal bindingness (based upon validity) is the default presumption, but does not exclude the possibility of mere moral bindingness.
10. This definition includes elements as presented by Levi-Faur (2010: 20).
11. This term will be explained in the below (see variables and maxims).
12. And of course these cases may be combined as often technological innovation leads to societal innovation.
13. 'Vertical' as in multi-level regulation; 'horizontal' as in multi-actor regulation; 'diagonal' as in problem-based cross-cutting horizontal and vertical (comp. Osofsky, 2009).
14. Within these 'institutional environments' different 'governance structures' define the state of play (i.e. organise transactions), such as contracts and firms on markets, cooperation within networks, administrative acts within hierarchy. Note that in institutional economics 'hierarchy' may refer to the governance structure of government but also to that of a firm (which is, internally, organised hierarchically) (See van Genugten, 2008).

15. Note that all three notions of legitimacy imply 'accountability', either democratically (as in the joint adages: 'no power without accountability' and 'no accountability without power'), in a meeting of supply and demand (through a barter involving an exchange of information), and in reciprocal activity (towards a shared/common objective).

16. Notions of political philosophy are clearly relevant (as is the concept of good legal governance), but only in as much captured within the legal framework (of the rule of law) – such as in 'democracy under the rule of law'.

17. The term 'likeliness' is used to depict that often, in practice, situations arise which do not allow for a ready or immediate judgment in terms of lawfulness – i.e. there is no consensus to that effect – and so a 'digital' legal qualification (i.e. lawful or unlawful) of a contested act or state of affairs is (for the time being) 'indeterminate' (and possibly subject to propositions and hypotheses, based upon doctrinal legal theory).

18. The importance of a good proxy may point at two different aspects: 1. the best description of the desired kind of behaviour (given that regulation is necessarily abstract and porous/underdetermined) – for example, 'establishment' as a forbidden industrial activity unless permitted (matching norm (output) and possible factual conduct (as outcome)); 2. the best behavioural mode considering the relevant public interest (given that this interest may be served by different kinds of behaviour) – for example, provisions prescribing targets versus prescribing use of certain means in curbing pollution by industrial emissions (matching outcomes in conduct with outcome in fostering the public interest).

19. Conduciveness is fostered by positive incentives, such as conviction (i.e. 'the right thing to do') or greater efficiency (i.e. less costs or more gains/profits for norm-addressees – e.g. tradable allowances) so as to align private behaviour to public interests.

20. Compare: 'Y shall not do X', as in 'X = killing or stealing' (categorical prohibition); 'Y shall not do X when Z, as in 'X = drive a car without lights on' and 'Z = after sunset' (hypothetical prohibition).

21. For example (1) by prohibiting pollution of 'type X', producers shift to or increase their pollution of 'type Y' – possibly even more detrimental to the environment as the relevant public interest. For example (2) negative side effects on an adjacent public interest: environmental norms which lead to practices that jeopardise safety in the workplace. Of course, trade-offs between different public interests are often unavoidable.

22. The 'smart bomb' metaphor describes a bomb smart enough to redirect itself. Mainly to be able to hit a moving target, but possibly also to ensure that no unnecessary ('collateral') harm is inflicted (i.e. to other objects/subjects than intended).

23. In turn this relates to the distinction between transaction costs and production costs (and effects), upon which this contribution shall not elaborate. Consider though, as an example, establishing or adjusting a chemical plant in accordance with new emission norms, to operate subsequently profitably also by selling redundant emission capacity.

24. The IEA (International Ergonomics Association) defines 'ergonomics' (or 'human factors') as 'the scientific discipline concerned with the understanding of interactions among humans and other elements of a system, and the

profession that applies theory, principles, data and methods to design in order to optimise human well-being and overall system performance. (IAE website [retrieved 27 February 2011]: http://iea.cc/.) The work of ergonomists is, again according to the IEA, concerned with 'the analysis of human-system interaction and the design of the system in order to optimise human well-being and overall system performance'. The analogy, of course, is the 'fit' between norm-addressees and regulation, and the ambition of optimally balancing 'norm-addressees well-being' and 'overall regulatory system performance'.

25. This point is also addressed in the 'Better Regulation'/'Smart Regulation' debate – see the references in the conclusion of this chapter.
26. Compare Sharpf's analyses on 'output-legitimacy' (Scharpf, 1999).
27. See endnote 14 on these 'institutional environments'. Furthermore, we will not go into the nature of this game in terms of being a '(non) zero-sum-game'.
28. So far published only as pre-publications on the LEGS website (http://www.utwente.nl/mb/legs/), but to be published in the near future (for reference see LEGS website)
29. To also write 'type' in italics is my personal choice, as it refers to a generalised perspective on the function-form relationship and the possibility of multiple use of the design (in cases where the same boundaries apply).
30. And not otherwise, such as, mean and lean, '*copy-paste*' strategies, with only minor amendments (on dates, names, objects ...). Thus existing legal artefacts are used as a design-form for other artefacts of a similar type. Merely applying a form(at) does not amount to design, especially if it lacks the reflection on methodological accurateness of this de facto design for the functionality of the specific object of creation.
31. Quoted almost literally (Ruiter, 2010c: 10).
32. The latter component is not relevant in categorical norms – at least not in terms of a legal constraint.
33. Creation implies the possibility of changing.
34. Consider, for example, how framing a conjunction of a (simultaneous) permission ('may do x') and an exemption ('may not do x'), which amounts to 'freedom' ('to do or not to do x'), may be regarded as a norm (as a strong or positive permission/exemption) or as absence of a norm (no prohibition and no command – a weak or negative permission) (von Wright, 1963: 86; Ruiter, 2010c: 2). Of course, such conjunctions may also be applied in the design of legal rights, according to the Hohfeld squares of legal relations.

References

Baldwin, R., 'Is better regulation smarter regulation?', *Public Law*, (2005), pp. 485–511.

Brownsword, R. *Rights, Regulation, and the Technological Revolution* (Oxford: Oxford University Press, 2008).

European Commission, Simplifying and improving the regulatory environment (*Action Plan on Better Regulation*), Com (2002) 278 final, Brussels, June 2002.

European Commission, *Better Regulation for Growth and Jobs in the European Union*, COM (2005) 97, Brussels, March 2005.

European Commission, *Smart Regulation in the European Union*, COM (2010) 543 final, Brussels, November 2010.

Genugten, M. L. van, *The Art of Alignment: Transaction Cost Economics and the Provision of Public Services at the Local level* (Dissertation, University of Twente, 2008).

Gunningham, N. and D. Sinclair, *Designing Smart Regulation*, OECD web publication, 1997. Available at http://www.oecd.org/dataoecd/18/39/33947759.pdf. Accessed on 25 September 2011.

Gunningham, N., P. Graboski and D. Sinclair, *Smart Regulation. Designing Environmental Policy* (Oxford: Oxford University Press, 1998).

Heldeweg, M. A., *Smart Rules & Regimes. Publiekrechtelijk(e) ontwerpen voor privatisering en technologische innovatie* (Elaborated inaugural lecture, Enschede, University of Twente 2010).

Heller, M., 'The Tragedy of the Anticommons', *Harvard Law Review*, vol. 1 (1998).

Klink, B. van, and C. Prins, *Law and Regulation. Scenarios for the Information Age* (Informatization Developments and the Public Sector Nr. 7), (Amsterdam: IOS Press, 2002).

Levi-Faur, D., 'Regulation & Regulatory Governance', *Jerusalem Papers in Regulation & Governance*, Working Paper No. 1, February 2010.

Mandelkern Group on Better Regulation, *Final Report (on Better Regulation)*, 13 November 2001. Available at http://ec.europa.eu/governance/better_regulation/documents/mandelkern_report. pdf. Accessed on 25 September 2011.

Murray, A. and C. Scott, 'Controlling the new media: hybrid responses to new forms of power', *Modern Law Review*, (2002), p. 491.

OECD, *From Red Tape to Smart Tape* (Paris: OECD, 2003).

OECD, *OECD Guiding Principles for Regulatory Quality and Performance* (OECD, 2005).

OECD, Economic Surveys, Netherlands (Paris: OECD, 2006)

Osofsky, H. M., 'Is climate change 'International'? Litigation's Diagonal Regulatory Role', *Virginia Journal of International Law*, vol. 29 (2009) no. 3, pp. 585–650.

Rittel, H. and M. Webber, 'Dilemmas in a general theory of planning', *Policy Sciences*, vol. 4 (Amsterdam: Elsevier Scientific Publishing Company, Inc., 1973). Reprinted in N. Cross (ed.), *Developments in Design Methodology* (Chichester: John Wiley & Sons, 1984).

Ruiter, D. W. P., *Institutional Legal Facts, Legal Powers and their Effects* (Dordrecht: Kluwer Academic Publishers, 1993).

Ruiter, D. W. P., *Legal Institutions* (Dordrecht: Kluwer Academic Publishers, 2001).

Ruiter, D. W. P., 'Four Legal Methodology Papers' (a. *A Methodology of Legal Design*; b. *A Methodology of Legal Institutional Design*; c. *A Methodology of Design of Legal Norms of Conduct*; d. *A Methodology of Design of Power Conferring Legal Norms*; e. *A Methodology of Legal Acts*), working papers, University of Twente, 2010 (a–e). Forthcoming (at http://www.utwente.nl/mb/legs/).

Ruiter, D. W. P., 'Calabresi and Melamed's entitlements: a Hohfeldian approach to 'The Cathedral', in M. L. van Genugten and M. Harmsen (eds), *De vorm behouden, verslag van een levenswerk door Dick W. P. Ruiter* (Enschede: University of Twente, 2008), pp. 61–87.

Scharpf, F.W., *Governing in Europe: Effective and Democratic?* (Oxford: Oxford University Press, 1999).

Scott, C., 'Regulatory governance and the challenge of constitutionalism', *EUI Working Papers* (RSCAS 2010/07), Robert Schuman Centre for advanced studies, Private regulation Series-02, 2010.

Somsen, H., 'Rechtvaardige en doelmatige regulering van medische biotechnologie: embryoselectie en biobanken', in H. Somsen, J. Bovenberg, and B. van Beers (eds), *Humane biotechnologie en recht* (Deventer: Preadviezen NJV, 2009).

Tiebout, C., 'A pure theory of local expenditure', *Journal of Political Economy*, vol. 64 (1956) no. 5, pp. 416–24.

Wetenschappelijke Raad voor het Regeringsbeleid (WRR), *Het borgen van het publiek belang*. Rapporten aan de regering nr. 56, 2000 (ISBN 901209058x).

Wetenschappelijke Raad voor het Regeringsbeleid (WRR) *Innovatie vernieuwd. Opening in viervoud.* WRR rapport nr. 80, Amsterdam University Press 2008 (the WRR website, see endnotes, also includes a summary in English: Innovation Renewed).

Wright, G. H. von, *Norm and Action* (London: Routledge and Kegan Paul, 1963).

4

Regulating Technological Innovation through Informal International Law: The Exercise of International Public Authority by Transnational Actors

Ramses A. Wessel

4.1 Introduction

The relationship between law, innovation and technology has been studied extensively. Almost by nature a warm relationship, it seems inherently contradictory as more law would imply less innovation. Even if we would point to the advantages of regulation for technological innovation, the institutional setting may be too complex to handle. In the words of Roger Brownsword and Han Somsen: 'In the best of all worlds, the regulatory environment will support and prioritise technological innovation that promises to strengthen the conditions that are essential for human social existence, and it will guard effectively against the abuse of and inherent risks presented by particular lines of technological development. In the real world, however, regulators have limited control over the priorities set by either market or military and there are severe restrictions on what nation states can individually control beyond their geographical borders.'[1] The present contribution purports to add to the existing body of literature by focusing on one specific phenomenon, which may complicate the institutional setting even further: informal international law as a tool to regulate technological innovation.

From the outset, international organisations have played a role in the international regulation of technology. In fact, some of the oldest international organisations were established exactly to regulate and facilitate international technological cooperation. Thus, for instance, the International Telecommunication Union (ITU) was established in

1865 and its objectives already hinted at the regulation of innovation: 'to promote the development of technical facilities and their most efficient operation with a view to improving the efficiency of telecommunication services, increasing their usefulness and making them, so far as possible, generally available to the public' (Art. 1.1.c of the ITU Constitution). Another example of an international organisation with a direct substantive link to technology is WIPO, the World Intellectual Property Organizaton. At the same time, although perhaps more indirectly, the international regulation of technology has become part and parcel of the tasks of organisations in the fields of for instance the environment, food, health or security, where we see novel applications of emerging technologies.[2]

The increasing influence of international organisations in general revealed that 'law-making is no longer the exclusive preserve of states'.[3] Indeed, international organisations are engaged in normative processes that, *de jure* or de facto, impact on states and even on individuals and businesses.[4] Decisions of international organisations are increasingly considered a source of international law,[5] and it is quite common to regard them in terms of international regulation or legislation.[6] As far as regular formal international organisations are concerned, their competence to take binding decisions vis-à-vis their member states is undisputed. They may even exercise sovereign powers, including executive, legislative and judicial powers.[7]

In addition, and apart from formal international organisations, an increasing number of other fora and networks have been recognised as playing a role in international or transnational normative processes. As José Alvarez notes, more and more technocratic international bodies 'appear to be engaging in legislative or regulatory activity in ways and for reasons that might be more readily explained by students of bureaucracy than by scholars of the traditional forms for making customary law or engaging in treaty-making; [t]hey also often engage in law-making by subterfuge'. Students of international relations and public administration pointed to the fact that the absence of a world government did not stand in the way of an 'emerging reality of global governance'. Recently, Jonathan Koppell sketched, both empirically and conceptually, the 'organisation of global rulemaking'. Even in the absence of a centralised global state, the population of Global Governance Organisations (GGOs) is not a completely atomised collection of entities. 'They interact, formally and informally, on a regular basis. In recent years, their programmes are more tied together, creating linkages that begin to weave a web of transnational rules and regulations'.[8] We now see a network of multiple GGOs consisting of a variety of governmental, non-governmental and hybrid organisations, which have as their main objective the crafting of rules and standards for worldwide application.[9]

The involvement of non-governmental actors in global rule making is far from new.[10] Even in the 'intergovernmental' ITU, private companies traditionally play an important role and some are even members of organs of the ITU (Art. 19 ITU Convention).[11] Nevertheless, we have increasingly become to realise that the global governance is in the hand of, what we term here, *informal* international bodies, which do not follow the traditional rules on international law making. In some issue areas, there is intense cooperation between state and non-state actors, such as in the regulation of the Internet by ICANN (the Internet Corporation for Assigned Names and Numbers). In some areas, states have even ceased to play a regulatory role, and transnational actors have taken over.[12] A prime example is the International Standardization Organization (ISO).[13]

While in most states the decisions of international organisations and bodies typically require implementation in the domestic legal order before they become valid legal norms, the density of the global governance web has caused some interplay between the normative processes at various levels. For EU member states (and their citizens), this can imply that the substantive origin of EU decisions (which usually enjoy direct effect in, and supremacy over, the domestic legal order) is to be found in another international body.[14] At the same time 'informal' rules often are adopted by regular international organisations, such as the World Trade Organization (WTO), which allows them to become part of 'formal' international decisions. However, informal decisions also may have an independent impact on domestic legal orders. The de facto impact of the – often quite technical – norms and the need for consistent interpretation[15] may thus set aside more sophisticated notions of the applicability of international norms in the domestic legal order.

Similar to the notion of multilevel governance as developed in political science and public administration, from a legal perspective the interactions between global, European and national regulatory spheres lead to the phenomenon of 'multilevel regulation'.[16] 'Regulation' is then defined in a broad sense, referring to the setting of rules, standards or principles that govern conduct by public and/or private actors. Whereas 'rules' are the most constraining and rigid, 'standards' leave a greater range of choice or discretion, while 'principles' are still more flexible, leaving scope to balance a number of (policy) considerations.[17]

In other words, 'regulation' refers to 'any instrument (legal or non-legal in its character, governmental or non-governmental in its source, direct or indirect in its operation, and so on) that is designed to channel behaviour'.[18]

In the following sections we will subsequently introduce the concept of informal international law making (section 4.2); analyse the importance of the 'exercise of public authority' by regulatory bodies (section 4.3); focus on a number of international bodies involved in the regulation of technology (section 4.4); and draw some conclusions on the relevance and the consequences of these forms of regulation (section 4.5).

4.2 Regulation through informal international law

The above analysis points to the recognition of norms that while being enacted *beyond* the state may nevertheless have an impact *within* the state. Indeed, domestic legal systems – traditionally, by definition, caught in national logic – increasingly recognise the influence of international and transnational regulation and law making on their development.[19] Legal scholars attempt to cope with the proliferation of international organisations and other entities contributing to extra-national normative processes.[20] Within this broader debate, a relatively new phenomenon has emerged: *informal* international law making. This type of law making is 'informal' in the sense that it dispenses with certain formalities traditionally linked to international law making. These formalities may have to do with *output, process* or the *actors involved*.[21] Pauwelyn defined informal international law making as: 'Cross-border cooperation between public authorities, with or without the participation of private actors and/or international organizations, in a forum other than a traditional international organization (process informality), and/or as between actors other than traditional diplomatic actors (such as regulators or agencies) (actor informality), and/or which does not result in a formal treaty or legally enforceable commitment (output informality)'.[22]

Informal international law making is novel in the sense that it goes beyond the 'law making by international organisations' debate[23] by focusing on other public authorities and normative outcomes, and differs from the more 'formal procedure-creating' approach of 'Global Administrative Law'.[24] At the same time, it shares some notions with the concept of 'multilevel regulation', both in terms of the actors involved and the effects that the normative output may have at different levels (global, regional (EU), domestic). Interestingly enough, informal international law making is based on the presumption that international cooperation, albeit less formal, falls within the remit of international law, on the ground that international law has, even traditionally, been defined with reference to its subjects (e.g. inter-state relations) rather than its object (be it subject matter or the particular form or type of output).[25] In that sense, informal international law making may manifest 'an impact-based conception of international law'.[26]

For the purpose of the present contribution it is important that 'informal international law' reveals that the regulation of technological innovation has moved from traditional intergovernmental settings to a complex web of regulatory bodies (*infra* section 4.3). While more forms present themselves, two types of international bodies in particular seem to play a role in informal international law making: *trans*-governmental networks and international agencies. Trans-governmental networks have been defined by Anne Marie Slaughter as 'informal institutions linking actors across national boundaries and carrying on various aspects of global governance in new and informal ways'.[27] These trans-governmental networks exhibit 'pattern[s] of regular and purposive relations among like government units working across the borders that divide countries from one another and that demarcate the 'domestic' from the 'international' sphere'.[28] They allow domestic officials to interact with their foreign counterparts directly, without much supervision by foreign offices or senior executives, and feature loose structures and peer-to-peer ties developed through frequent interaction.[29] The networks are composed of national government officials, either appointed by elected officials or directly elected themselves, and they may be among judges, legislators or regulators.[30] According to Kanishka Jayasuriya, these new regulatory forms have three main features: (1) they are governed by networks of state agencies acting as independent actors rather than on behalf of the state but; (2) they lay down standards and general regulatory principles instead of strict rules; and (3) they frequently contribute to the emergence of a system of decentralised enforcement or the regulation of self-regulation.[31] A trans-governmental regulatory network is basically cooperation between regulatory authorities of different countries.

In addition, regulation may be in the hands of what we have coined 'international agencies': international bodies that are not based on an international agreement, nor on bottom-up cooperation between national regulators, but on a decision by an international organisation.[32] These bodies have also been referred to as 'transnational administrative networks' (TANs) and their composition may differ substantially from that of the 'mother' organisation, for instance through the participation of experts from industry and from NGOs.[33] According to some observers, these new international entities even outnumber the conventional organisations.[34] International regulatory cooperation is often conducted between these non-conventional international bodies[35] and it is not unusual for international agencies to engage in international norm-setting. Here also, the tendency towards functional specialisation because of the technical expertise required in many areas may be a reason for the proliferation of such bodies and for their interaction with other international organisations and agencies, which sometimes leads to the creation of common bodies. International (regulatory) cooperation is often conducted between these non-conventional international bodies.[36] Whereas traditional international organisations are established by an agreement between states, in which their control over the organisation and the division of powers is laid down,[37] the link between newly created international bodies and the states that established the parent organisation is less clear. As one observer holds, this 'demonstrates how the entity's will does not simply express the sum of the member states' positions, but reformulates them at a higher level of complexity, assigning decision-making power to different subjects, especially to the international institutions that promoted the establishment of the new organization'.[38]

This leads to a large number of potential fora involved in informal international law making. First of all, what has been set out above already indicates that governance, and by the same token regulation, has become a multi-actor game; apart from intergovernmental organisations, non-governmental and transnational actors are playing an increasing role in global governance.[39] National agencies thus participate in global (or regional) regulatory networks as largely independent, autonomous actors and are, in turn, often required to implement international regulations or agreements adopted in the context of these networks at the national level.[40] As early as a decade ago, Anne-Marie Slaughter termed this phenomenon the 'nationalization of international law',[41] and because of the fast developments in technology and the specific expertise required in that sector, the phenomenon has presented itself clearly in particular in relation to the regulation of technology. As one observer held, this is 'governance by technical necessity'.[42]

4.3   The exercise of public authority through regulatory activities

The potential list of international bodies that are somehow involved in rule making is thus quite extensive. In our quest to look for regulatory bodies that are somehow involved in informal international law-making, it may be wise to follow Jonathan Koppell's suggestion to focus on those organisations that are 'actively engaged in attempts to order the behaviour of other actors on a global scale'.[43] Only organisations devoted to normative, rule creating and rule supervisory activities would thus be GGOs.

However, as we are mainly interested in the 'public' dimension of regulation, we wish to be even more precise. Following the notion that 'governance' is about creating (public) order,[44] a good starting point may be to raise the question whether 'public authority' is exercised when we look at rule making. This notion was recently studied within the framework of a Max Planck project on the 'Exercise of International Public Authority'.[45] Large parts of international cooperation (including some of the forms mentioned above) could be considered as merely affecting the private legal relationships between actors. We would argue that the 'public' dimension is essential whenever we wish to study new forms of lawmaking; irrespective perhaps of the process, the actors or the instruments used. Armin von Bogdandy, Philipp Dann and Matthias Goldman define the 'exercise of international public authority' in the following terms: 'any kind of governance activity by international institutions, be it administrative or intergovernmental, should be considered as an exercise of international public authority *if* it determines individuals, private associations, enterprises, states, or other public institutions'.[46] 'Authority' is defined as 'the legal capacity to determine others and to reduce their freedom, i.e. to unilaterally shape their legal or factual situation'.[47] Also important is the fact that the determination may or may be not legally binding: 'It is binding if an act *modifies the legal situation* of a different legal subject without its consent. A modification takes place if a subsequent action which contravenes that act is illegal'.[48] The authors believe that this concept enables the identification of all those governance phenomena, which public lawyers should study. Irrespective of its focus on 'governance activity by international institutions', we feel that this definition may also be applied to the informal fora addressed in the present contribution and could thus be applicable to all GGOs, including the ones involved in the regulation of technological innovation.

Whereas the authors convincingly argue that that the capacity to determine another legal subject can also occur through a non-binding act, which only conditions another legal subject, we would limit the concept of 'regulation' to activities that do indeed modify the legal situation of a different legal subject. At the same time we wish to rule out pure private authority exercised by transnational or international bodies (as well as companies). The 'publicness' of the international act therefore seems important and may be the most difficult element to establish. After all – as also noted by Von Bogdandy, Dann and Goldmann – it would be too easy to relate the 'publicness' of a legal act to an existing legal basis for the authority. We cannot exclude that (de facto) public authority is exercised by non-governmental or hybrid international institutions, which may only be indirectly based on a state's 'consent to be bound'.

The concept would thus cover not only (many, but not all) decisions by formal international organisations, but also forms of law making that because of the nature of the body, the process or the instrument may be more informal.

4.4   Examples of regulation of technological innovation through informal international law

In order to limit the size of this contribution, we will focus on bodies with activities that are directly relevant for technological sectors, more precisely the Internet sector. The Internet sector offers a number of examples of technological regulation and may serve as a good illustration of informal international law making.[49] This leaves out other bodies in the technical arena, such as ISO, as well as all bodies with powers to regulate technological innovation in areas such as the environment, food, health or security. We will also leave out the formal activities of traditional international organisations, as these are well described by others.[50] The descriptions below merely serve as illustrations of regulation of technological innovation through informal international law making. Obviously, a broader scope would reveal a larger number of involved regulatory bodies. The main purpose will be to find out when we can argue that 'international public authority' is exercised.

ICANN

ICANN is a non-profit corporation, with the mission of coordinating the global Internet's systems of unique identifiers. It coordinates the allocation and assignment of the three sets of unique identifiers for the Internet: domain names; Internet protocol (IP) addresses and autonomous system (AS) numbers; and protocol port and parameter numbers. ICANN is a non-profit corporation under Californian law and therefore is a striking example of a body that despite not being an international organisation seem to govern an entire technical sector on a global scale.

ICANN thus defies the traditional foundations of international law making: its main members are private corporations (with national governments in an advisory role only), it has no international legal status and it is not based on an international agreement in which its competences are laid down and restricted. Thus, formally, ICANN does not exist in international law. Yet, as argued by one observer, 'ICANN establishes rules which are of greater importance than most acts of international organizations and they are more widely and more strictly accepted and respected than binding decisions of most international organizations. One could make the argument that ICANN decisions are more authorative than those of the UN Security Council in the sense that ICANN decisions are less frequently violated'.[51] The reason is simple: ICANN's rules are necessary for the operation of the Internet.

The private law origin of ICANN is reflected in the composition of its main decision-making body, the Governing Board, which draws its members from interested organisations and groups. Governments do have an influence through one of the advisory bodies only, the Governmental Advisory Committee (GAC). The GAC is composed of representatives of (102) state governments, public authorities and (14) representatives of international organisations (such as the ITU and the WIPO). Since 2002 (following the terrorist attacks of 2001), the GAC's advises are duly taken into account by the Board of Governors (see Art. 1, sec. 2.11 Bylaws of 2002). The GAC has its own governance structure, secretariat and decision-making procedures and seems to have become an 'intergovernmental organisation within a non-governmental organisation'.

Another element pointing to its 'informal law' status concerns the 'output'. ICANN does not regulate on the basis of binding decisions. Rather, it concludes contracts with the registries in charge of the administration of Internet 'top-level domains' (TLDs). However, given the fact that Internet access is dependent on having a TLD name (such as .eu), one may argue that this comes close to 'de facto' bindingness. Indeed 'It seems quite logical that the uniformity of the rules is best guaranteed by a single "legislator"'.[52] It is this argument that seems to form the source of many more examples of the regulation of technology. Despite its informal, non-governmental, nature, ICANN fulfils a public task. It administers a scarce common good and decides on its assignment. In that sense it indeed can be said to exercise public authority.[53]

The Internet Governance Forum (IGF)

The 2006 World Summit on the Information Society (WSIS) led to the establishment of the IGF, with a view to better understanding issues related to internet governance and to promoting dialogue among stake-holders in an open and inclusive manner. The mandate of the Forum is laid down in Paragraph 72 of the Tunis Agenda adopted by the WSIS, which was endorsed by the UN General Assembly in its Resolution 60/252.

Unlike ICANN, the IGF allows for more groups to participate in meet-ings: governments, the private sector, civil society, intergovernmental and other international organisations. In the 2010 meeting (in Vilnius), 1451 people participated (a total of around 2000 persons were present). The breakdown of participants shows that all the major stakeholder groups were represented almost equally, with 21 per cent of partici-pants coming from civil society, 23 per cent from the private sector, 24 per cent comprising government representatives and 22 per cent made up of technical and academic communities. Institutionalisation took place on the basis of the creation of a de facto secretariat, the Multi-stakeholder Advisory Group (MAG). This MAG has 56 members, which are nominated by the different stakeholder groups taking into account geographical and gender balance. The MAG prepares the IGF meetings and meets three times per year; it is physically located within the UN Offices in Geneva.

Apart from the Chairman Submissions that are issued at the end of every meeting, IGF meetings have no formal binding output. Nevertheless, the IGF is believed to affect decisions that are taken elsewhere. Thus, the work of the IGF has been reflected in Ministerial Declarations of the Council of Europe and the OECD.[54]

While the IGF most certainly influences the political as well as technical governance of the Internet, it would be hard to argue that it exercises public authority itself. It does play its role in the regulation of the Internet and may in that sense have a public task. It does, however, seem to lack 'the legal capacity to *determine* others and to reduce their freedom, that is, to unilaterally shape their legal or factual situation'.

The Internet Engineering Task Force (IETF) and the Internet Society (ISOC)

ISOC is an organisation network for the groups responsible for Internet infrastructure standards, including the IETF. The latter is the principal body engaged in the development of new Internet standard specifica-tions. Being a large open international community of network designers, operators, vendors and researchers, IETF is responsible for the resolution of all short- and mid-range protocol and architectural issues required to make the Internet function effectively. IETF is a network, formally established by IAB (Internet Architecture Board). It is not a corporation and it lacks a definite legal status. It has no board of directors, no official members and no dues. ISOC is an independent international non-profit organisation, established in 1992 with the purpose of providing institutional framework and financial support for IETF, but it later expanded its objectives. ISOC is a corporation, incorporated under the District of Columbia Non-Profit Corporation Act. Its responsibilities are provided for in RFC 1602 (Revision 2 of The Internet Standards Process), a constitutive instrument that was adopted in 1992 and was later revised.

There is no membership in the IETF. Anyone can register for and attend any meeting. The closest thing there is to being an IETF member is being on the IETF or one of the Working Groups' mailing lists. The usual participants are designers, operators, vendors and researchers concerned with the evolution of the Internet. Government representatives can participate in the process; however their participation is at the same level as that of any private individual or expert. They are not accorded any special treatment; on the contrary, they only form a part of a large Internet community. The membership of ISOC is more structured – it is open to individuals and organisations. Today, ISOC's community has more than 26,000 individual members. Groups of people who live in the same area or share an interest in specific issues can form an ISOC Chapter. ISOC's Organisation Members include corporations; non-profit, trade, and professional organisations; foundations; educational institutions; government agencies; and other national and international organisations.

The Internet Standards Process starts at the IETF. A specification undergoes a period of development and several iterations of review by the Internet community and revision based upon experience. The standards developed through the IETF are considered by the Internet Engineering Steering Group, with appeal to the IAB, and promulgated by the Internet Society as international standards. Typically, a standards action is initiated by a recommendation to the appropriate IEFT Area Director by the individual or group that is responsible for the specification, usually an IETF Working Group (WG). WGs cooperate through the mailing lists. An important fact is that there is no formal voting in a WG. The general rule on disputed topics is that the WG has to come to 'rough consensus', meaning that a very large majority of those who care must agree.

Output takes different forms: proposed standards, draft standards, internet standards, best current practices documents, informational documents, experimental documents and historical documents. The

Internet Standards Process deals with protocols, procedures, and conventions that are used in or by the Internet, whether or not they are part of the TCP/IP protocol suite. The effect of Internet standards is not binding per se, but the purpose of the Internet standards making process is to get consent from end-users and their affirmation of the standard. This will result in actual use of the standards and therefore, a more unified and open use of the Internet. In effect, the Internet Standards Process has a very concrete and formal output and its standards are widely used by the Internet community.

This feature reveals the public authoritative nature of the process. It is hardly possibly not to accept the standards, which leads to an effective international regulation of this area through different, informal, means.

The Global Cybersecurity Agenda (GCA) of the ITU

GCA is a framework for international cooperation aimed at enhancing confidence and security in the information society. It was launched in 2007 by the ITU Secretary-General. Cybersecurity refers to protection against unauthorised access, manipulation and destruction of critical resources. The main problem is the lack of international harmonisation regarding cybercrime legislation. ITU's idea with GCA is that the strategy for a solution must identify those existing national and regional initiatives, in order to work effectively with all relevant players and to identify common priorities.

All members of the ITU – 191 member states and 700 sector members – can participate in discussion and initiatives of the GCA. The decision-making process depends on the decision taken. For example, recommendations are issued on the basis of a consensus of all participants. On the other hand, toolkits ('model laws') are prepared by lawyers and not by state representatives. Except from the formal establishment of the initiative there is no 'output' as such, the objective being to influence the practice worldwide. With its cybercrime legislation resources and material, the GCA under the ITU aims to assist countries in understanding the legal aspects of cybersecurity in order to move towards a harmonising of legal frameworks. Apart from many key security Recommendations, ITU has developed overview security requirements, security guidelines for protocol authors, security specifications for IP-based systems, guidance on how to identify cyber threats and countermeasures to mitigate risks. One of the most important security standards in use today is X.509, an ITU-developed Recommendation for electronic authentication over public networks.

Recently, ITU-T X.1205 'Overview of Cybersecurity' was approved. It provides a definition of cybersecurity and taxonomy of security threats. It discusses the nature of cybersecurity environment and risks, possible network protection strategies, secure communications techniques and network survivability.

Irrespective of their influence, the decisions taken do not have a binding force for the Members of the GCA. Again, however, one may argue that once the adopted recommendations in effect regulate a particular area for instance, by excluding other possibilities the GCA is exercising international public authority. Given the subject matter, however, this effect may only occur once market players or governments decide on a mandatory use of the adopted standards.

4.5   Consequences of the regulation of technological innovation through informal international law making and suggestions for further research

There is nothing new in arguing that 'regulation beyond the state' seems to have replaced traditional forms of legal governance. In legal science, however, the impact of this development is much larger than in, for instance, public administration. Lawyers tend to work with 'legal systems' that are neatly separated and have their own source of norms. While the debate on 'multilevel governance' can said to have taken place within the academic disciplines of political science and public administration, the phenomenon of 'multilevel regulation' challenges the very foundations of law itself.

The notion of 'informal international law making' aims to find a way out of the tension between traditional legal science (with its focus on 'sources', 'jurisdiction' and 'competences') and the factual reality of norms being enacted by actors and through procedures that are unfamiliar to the traditional lawyer. Yet, as the cases on the regulation of the Internet show, the impact – even in a legal sense – of these norms may be larger and more widespread than formal treaty law or decisions by international intergovernmental organisations.

While the transfer of competences to formal international organisations is a careful process guided by strict rules and principles (such as the 'principle of the attribution of powers'[55]), competences seem to have been transferred to or created by more informal fora in a parallel process. Again, this is not new,[56] but the extent to which large parts of society now seem to be regulated in 'informal' ways has triggered a debate on the consequences (in terms of legitimacy and accountability, or more generally upholding the rule of law) and possible solutions (ranging from the introduction of constitutional principles at the global level, the development of global administrative law, or the acceptance of the plurality of legal orders and the fragmentation of international law).[57] These responses underline that we may indeed have to rethink certain traditional aspects of international law.

The regulation of technology is a prime example of an area which is already largely outside the direct influence of the traditional lawmakers, the states. At the same time, we have seen that in most cases we are not talking about small and select groups of actors. Many stakeholders are involved and the institutionalisation has shown a dynamic that is similar to traditional international organisations. Moreover, this is not about the private sphere of companies; in many cases international public authority is exercised. It is clear that there is no way back and that 'global governance' has developed either in the shadow of existing arrangements or simply 'bottom up' through cooperation between national regulators. The reason is obvious. The regulation of technology can only be done by experts ('governance by technical necessity').

Legal science is only at the beginning of accepting the reality of this development. At the same time this offers an opportunity to rethink the relationship between law, innovation and technology. More research on 'informal international law making' may assist in providing the necessary empirical data and conceptual notions to square the contradiction presented at the beginning of this contribution.

Notes

1. See Brownsword, R. and H. Somsen, 'Law, innovation and technology: before we fast forward – a forum for debate', *Law, Innovation and Technology* (2009), no. 1, pp. 1–73.
2. Ibid., p. 1.
3. Boyle, A. and C. Chinkin, *The Making of International Law* (Oxford: Oxford University Press, 2007), at vii. See for a non-legal approach: M. J. Warning, *Transnational Public Governance: Networks, Law and Legitimacy* (Basingstoke: Palgrave Macmillar, 2009).
4. See Føllesdal, A., R. A. Wessel and J. Wouters (eds), *Multilevel Regulation and the EU: The Interplay between Global, European and National Normative Processes* (Leiden and Boston: Martinus Nijhoff Publishers, 2008).
5. See also Dekker, I. F. and R. A. Wessel, *Governance by International Organisations: Rethinking the Source and Normative Force of International Decisions*, I. F. Dekker and W. G Werner (eds), *Governance and International Legal Theory* (Leiden and Boston: Martinus Nijhoff Publishers, 2004), pp. 215–36.
6. Whereas 'regulation' is the more comprehensive term used in this contribution, 'legislation' has a more narrow connotation, as 'legislative power' has

been said to have three characteristics: (1) a written articulation of rules that (2) have legally binding effect as such and (3) have been promulgated by a process to which express authority has been delegated *a priori* to make binding rules without affirmative *a posteriori* assent to those rules by those bound. See Oxman, B., 'The international commons, the international public interest and new modes of international lawmaking', in J. Delbrück (ed.), *New Trends in International Lawmaking – International 'Legislation' in the Public Interest* (Berlin: Ducker & Humblot, 1996), at pp. 28–30. Cf. Also the 'Comments' by T. Stein and C. Schreuder in the same volume. An even more distinguishing element, perhaps, is that such rules imply future application to an indeterminate number of cases and situations. See A. J. J. de Hoogh, *Attribution or Delegation of (Legislative) Power by the Security Council?*, Yearbook of International Peace Operations, 2001.

7. See Sarooshi, D., *International Organizations and their Exercise of Sovereign Powers* (Oxford: Oxford University Press, 2005).

8. Koppell, J. G. S., *World Rule. Accountability, Legitimacy, and the Design of Global Governance* (Chicago/London: The University of Chicago Press, 2010).

9. See for a recent survey of examples O. Dilling, M. Herberg and G. Winter (eds), *Transnational Administrative Rule-Making: Performance, Legal Effects and Legitimacy* (Oxford: Hart Publishing, 2011).

10. In fact, the current attention for transnationalism may remind us of the dominant debate in international relations theory in the beginning of the 1970s. See for instance R. Keohane and J. Nye, 'Transnational relations and world politics', *International Organization* (1971), p. 329. See also Chr. Tietje, 'History of transnational administrative networks', in Dilling, Herberg and Winter, op. cit., pp. 23–37.

11. Cf. Jacobson, H. K., 'ITU: A Potpourri of Bureaucrats and Industrialists', in R. W. Cox and H. K. Jacobson, *The Anatomy of Influence, Decision-Making in International Organizations* (New Haven: Yale University Press, 1973).

12. *See* Hartwich, M., 'ICANN – Governance by Technical Necessity', in A. Von Bogdandy, R. Wolfrum, J. Von Bernsdorff, Ph. Dann and M. Goldmann (eds), *The Exercise of Public Authority by International Institutions: Advancing International Institutional Law* (Heidelberg, etc.: Springer, 2010), pp. 575–605.

13. See Hall, R. B. and Th. J. Biersteker (eds), *The Emergence of Private Authority in Global Governance* (Cambridge: Cambridge University Press, 2002).

14. For a survey of the relations between the EU and other international organizations, see generally Hoffmeister, F., 'Outsider or frontrunner? Recent developments under international and European Law on the status of the European Union in international organizations and treaty bodies', *Common Market Law Review*, (2007), pp. 41–68. See also Chiti, E. and R. A. Wessel, 'The emergence of international agencies in the global administrative space: autonomous actors or state servants?', in White, N. and R. Collins (eds), *International Organizations and the Idea of Autonomy* (London: Routledge, 2011).

15. Cf. Cottier, Th., 'A theory of direct effect in global law', in Von Bogdandy, A. et al. (eds), *European Integration and International Co-ordination: Studies in Transnational Economic Law in honour of Claus Dieter Ehlermann* (The Hague: Kluwer Law International, 2001), at 109–10 (discussing the impact of the doctrine of consistent interpretation in relation to the domestic effect of WTO law).

16. See Føllesdal, Wessel and Wouters, op. cit. Cf. also N. Chowdhury and R. A. Wessel, 'Conceptualising multilevel regulation: a legal translation of multilevel governance?', unpublished paper, forthcoming 2012.

17. Wessel, R. A. anc J. Wouters, 'The phenomenon of multilevel regulation: interactions between global, EU and National Regulatory Spheres', in Føllesdal, Wessel and Wouters, op. cit., pp. 9–47, at 11–12.

18. Brownsword and Somsen, op. cit., at 8. This description comes close to the widely accepted definition by Julia Black: 'the sustained and focused attempt to alter the behaviour of others according to standards or goals with the intention of producing a broadly identified outcome or outcomes, which may involve mechanisms of standard-setting, information-gathering and behaviour-modification.' J. Black, 'What is regulatory innovation?' in J. Black, M. Lodge and M. Thatcher (eds), *Regulatory Innovation* (Cheltenham: Edward Elgar, 2005), p. 11.

19. See generally Nijman, J. and P. A. Nollkaemper, *New Perspectives on the Divide between National & International Law* (Oxford: Oxford University Press, 2007).

20. See Jayasuriya, K., 'Globalization, law, and the transformation of sovereignty: the emergence of global regulatory governance', *Indiana Journal of Global Legal Studies* (1999), p. 425. In his book *International Organizations as Law-makers* (op. cit.), José Alvarez reveals that the role of international organisations in law making not only increased, but also that international law is not always well equipped to handle this development. Cf. Sarooshi, D., op. cit. For earlier examples, see J. Delbrück, op. cit. On the development of the (sub-)discipline of the law of international organizations in general, see J. Klabbers, 'The life and times of the law of international organizations', *Nordic Journal of International Law* (2001), pp. 287–317.

21. See Pauwelyn, J. 'Informal international law-making: mapping the action and testing concepts of accountability and effectiveness', in J. Pauwelyn, R. A. Wessel and J. Wouters (eds), *Informal International Lawmaking: Mapping the Action and Testing Concepts of Accountability and Effectiveness* (Oxford: Oxford University Press, 2012, forthcoming).

22. Ibid.

23. Alvarez, op. cit.

24. See Kingsbury, B. and L. Casini, 'Global administrative law dimensions of international organizations law', *International Organizations Law Review*, (2009) no. 2, pp. 319–56, as well as other contributions to the special issue of *IOLR* on *Global Administrative Law in the Operations of International Organizations*. Earlier: N. Krisch and B. Kingsbury, 'Introduction: global governance and global administrative law in the international legal order', *European Journal of International Law* (2006); as well as B. Kingsbury, N. Krisch and R. Steward, 'The emergence of global administrative law' *Law and Contemporary Problems* (2005), pp. 15–61, at p. 29; Harlow, C., 'Global administrative law: the quest for principles and values' *EJIL* (2006), pp. 197–214; B. Kinsgbury, 'The concept of law in global administrative law', *EJIL* (2009), pp. 23–57.

25. Pauwelyn, op. cit. Cf. also D. W. P. Ruiter and R. A. Wessel, 'The legal nature of informal international law: a legal theoretical exercise', in Pauwelyn, Wessel and Wouters, op. cit.

26. D'Aspremont, J., 'Informal International Public Policy Making: From a Pluralisation of International Norm-Making Processes to a Pluralisation of our Concept of International Law', paper presented at the workshop Informal International Public Policy Making, Geneva, 24–25 June 2010.
27. Slaughter, A. M. and D. Zaring, 'Networking goes international: an update', *Annual Review of Law and Social Science* (2006), p. 215.
28. Slaughter, A. M., *A New World Order* (Princeton University Press, 2004), p. 14.
29. Slaughter and Zaring, op. cit., p. 215; Raustiala, K., 'The architecture of international cooperation: Transgovernmental networks and the future of international law', *Virginia Journal of International Law* (2002–2003); Risse-Kappen, T., 'Introduction', in Risse-Kappen, T. (ed.), *Transnational Relations Back In*, 1995.
30. Slaughter, op. cit., pp. 3–4.
31. See Jayasuriya, op. cit., at 453. On the regulation of self-regulation in particular, see generally G. Teubner, 'Substantive and reflexive elements in modern law', *Law & Society*, (1983). Elements of this development are also addressed by Anne-Marie Slaughter, op. cit. Slaughter seems to use the term 'transgovernmental networks' to point to what we would call informal international law-making. Slaughter, op. cit., Chapter 6.
32. Chiti, E. and R. A. Wessel, op. cit.
33. Oeter, S., 'The openness of international organisations for transnational public rule-making', in Dilling, Herberg and Winter, op. cit., pp. 236–52 at 240.
34. See Shanks, C., H. K. Jacobson and J. H. Kaplan, 'Inertia and change in the constellation of international governmental organizations, 1981–1992', *International Organization*, (1996), pp. 593.
35. Cf. Tietje, C., 'Global governance and inter-agency cooperation in international economic law', *Journal of World Trade* (2002), p. 501.
36. Ibid., p. 501.
37. On the different dimensions of the relationship between states and international organizations cf. D. Sarooshi, op. cit.
38. Martini, C., 'States' control over new international organization', *Global Jurist Advances* (2006), pp. 1–25, at 25.
39. Anne-Marie Slaughter regards these networks as a better way of world governance than the traditional state-centric approach. See Slaughter, op. cit.
40. See Jayasuriya, op. cit., at 440. See also S. Picciotto, *The Regulatory Criss-Cross: Interaction Between Jurisdictions and the Construction of Global Regulatory Networks*, in W. Bratton et al. (eds), *International Regulatory Competition and Coordination: Perspectives on Economic Regulation in Europe and the United States*, 1996.
41. Anne-Marie Slaughter, 'The real new world order', *Foreign Affairs* (1997), p. 192.
42. Hartwich, op. cit.
43. Koppell, op. cit., at 77.
44. For example: Peters, B. G., 'Introducing the topic', in B. G. Peters and D. J. Savoie (eds), *Governance in a Changing Environment* (Montreal: MacGill-Queens University Press, 1995).
45. See Von Bogdandy et al., op. cit. See in the same volume also M. Goldmann, 'Inside relative normativity: from sources to standards instruments for

the exercise of international public authority', pp. 661–711; and A. Von Bogdandy, P. Dann and M. Goldmann, 'Developing the publicness of public international law: towards a legal framework for global governance activities', pp. 3–32.

46. Ibid., at 5.
47. Ibid., at 11.
48. Ibid., at 11–2.
49. Cf. also von Bernstorff, J., 'Democratic global Internet regulation? Governance networks, international law and the shadow of hegemony', *European Law Journal*, (2003), pp. 511–26.
50. Many examples are drawn from the case studies in the 'Informal International Law-Making' project (IN-LAW; see www.informallaw.org). The case studies used here were done by Ana Berdajs at the Graduate Institute in Geneva and credit is due to her work.
51. Hartwig, op. cit., p. 576.
52. Ibid., at 591.
53. Ibid., at 603. Cf. also B. Carotti and L. Cassini, 'Complex governance forms: hybrid, multilevel informal', in S. Cassese et al. (eds), *Global Administrative Law*, 2008. Available at: http://www.iilj.org/GAL/documents/GALCasebook. pdf; visited on 22 September 2011).
54. Based on the data retrieved in the IN-LAW project by Ana Berdajs. See also Note of the UN Secretary-General, Prepared by UN Department of Economic and Social Affairs (May 2010). Available at: http://unpan1.un.org/ intradoc/groups/public/documents/un-dpadm/unpan039074.pdf; visited on 22 September 2011.
55. See Schermers, H. G and N. M. Blokker, *International Institutional Law: Unity in Diversity* (Boston and Leiden: Martinus Nijhoff Publishers, 2003), p. 155: 'A rule of thumb is that, while states are free to act as long as this is in accordance with international law [...], international organizations are competent only as far as powers have been attributed to them by the member states. [...] International organizations may not generate their own powers. They are not competent to determine their own competence'.
56. Cf. the *Lex Mercatoria* governing transnational trade. See for instance L. M. Friedman, 'Erewhon: the coming global legal order', *Stanford Journal of International Law*, 92001), pp. 347–59.
57. See more extensively on this issue: R. A. Wessel, 'Reconsidering the relationship between international law and EU law: Towards a content-based approach?' in E. Cannizzaro, P. Palchetti and R. A. Wessel, *International Law as Law of the European Union* (Boston and Leiden: Martinus Nijhoff Publishers, 2011) (forthcoming).

Part II
Specific Exemplary Areas and Connected Issues of Regulating Technological Innovation

5

The Unlikely Emergence of Next Generation Networks in the Light of Prevailing Telecom Regulation: Instigating a decision supporting framework for stimulating network innovation (especially in telecommunications) based on first and second mover theory under network effects

Lesley C. P. Broos

5.1   Introduction

The current far-reaching European legislative reform of the telecommunications sector has led to a lively debate in the media about (1) the balance between consumer protection on the one hand and entrepreneurial freedom of telecommunication companies (providers) on the other hand (e.g. the discussions about the maximal minimum duration of telecom subscriptions for consumers and the discussions about network neutrality) and (2) what the impact of the new legislation would be on the innovativeness of the industry, for example: will mandated access to next generation networks stimulate or hamper investment by providers? The first debate is mainly political and will not be discussed here. This chapter focuses on the latter, which is mainly the result of insufficient scientific insights into the relation between regulation and innovation. Regarding this 'insufficient insight', Irwin and Vergragt (1989) already stated more than 20 years ago:

> Regulation has rarely been considered as a positive means of technical control e.g. through stimulating new forms of technological

response rather than simply restricting the operation of the market-place. The whole issue of regulation, therefore, has been conceptualized as a post-innovation check on undesired side-effects rather than as a tool for directing technology towards socially desirable ends.

In addition, Lundvall (1992) strongly questions the methodological basis of the rare literature trying to create such a 'tool'. Given the current debate as mentioned above, not much progress seems to have been made since then.

This chapter aims to shed new light on the possibilities of pacing innovation by means of regulation. More concrete, it will focus on meso-level (industry) discussions about stimulating the deployment of so-called next generation networks (NGN) for telecommunication. According to the Standardisation Sector of the ITU (2004), an NGN is a packet-based network able to provide telecommunications services and to make use of multiple broadband, QoS-enabled transport technologies and in which service-related functions are independent from underlying transport-related technologies. NGNs offer unrestricted access by users to different service providers and support generalised mobility which will allow consistent and ubiquitous provision of services to users.[1] However, in this chapter, we consider NGNs as a more relative label for networks based on new technological possibilities in comparison with already existing networks, which we will name 'same generation networks' (SGNs).

Especially in the telecommunications industry, the creation and exploitation of first and second mover advantages (following industrial organisation theory and discussed shortly below sub II) appear to be quite central in the strategy of telecom providers. So, influencing providers' strategies in a more innovative direction by means of regulation, could be realised by changing institutional factors determining the existence and strength of first and second mover advantages. However, some specific network industry-typical characteristics ('network effects', discussed later in the chapter) must be taken into account applying this innovation timing advantages theory.

Given a market in which a telecom provider with sustainable market power (a so-called SMP operator or incumbent) is present with a telecommunications network (based on e.g. cable or xDSL) and servicing end users already, we will consider the institutional influence of some regulatory practices on the attractiveness of several strategic options for a *new entrant* planning to enter that same marketplace.[2] Hereto, common regulatory practices in telecommunications will be confronted with innovation timing advantages and network effects. For this purpose, we assume three strategic options for the new entrant aiming at offering tele-communication services to end-users: (1) Creating a SGN, for example a second or third UMTS-network, or deploying an optical fibre network next to already existing similar optical fibre networks, (2) offering services making use of the incumbent's network ('service based competition' (SBC)) and (3) creating an NGN, for example deploying a fibre network in a region where until then only ADSL or cable coverage existed. For the purpose of this chapter, we consider SGN as the least innovative strategy. SBC might lead to more innovative *services* being available on the network involved (which might be an NGN already), but as this chapter focuses on innovation of the physical network itself, NGN is considered as the most innovative new entrance strategy. The question thus is, how to regulate new entrants in the direction of NGNs instead of SGNs or SBCs.

By exploring (1) network industry characteristics, (2) innovation timing (dis)advantages in telecoms based on first and second mover theory, and (3) institutional factors influencing these (dis)advantages, a decision supporting framework for stimulating network innovation (more specifically NGN development) will be introduced and some research directions for further improvement of this framework will be suggested.

This chapter is structured as follows. In the second section we will illustrate the major role of first and second mover advantages in the telecommunications industry. In the third section we will explain the concept of network effects, which have a serious impact on corporate strategy as well. In the fourth section, common regulatory practices in telecommunications will be confronted with innovation timing advantages and network effects to analyse the institutional influence of several regulatory practices on the attractiveness of new entrants' strategic options. Finally, in the last section some conclusions will be drawn about the relation between prevailing and intended telecom regulation versus the deployment of NGNs, and about the value and suggestions for further improvement of the instigated framework.

5.2   Innovation timing advantages

In this section we will illustrate the relevance of first and second mover advantages – for which Lieberman and Montgomery (1987) provided a unified conceptual framework as well as a critical assessment in a prize-winning paper – in the strategy of telecom providers. It goes without saying that the presence and strength of first and second mover advantages in telecom markets can differ according to regional circumstances (including the maturity of telecommunication facilities) and regulatory history in a certain region. Lieberman and Montgomery identified thee sources of first-mover advantages (FMAs) which may be enjoyed by firms which are – due to initial asymmetries – able to introduce innovative services in an early stage:[3] technological leadership, pre-emption of assets and buyer switching costs.

Technological leadership may result in lower production costs per unit because of pole position on the learning and experience curve. Furthermore, technological leadership might lead to more success in patent or R&D races, where advances in product or process technology are a function of R&D expenditures. As the importance of patented communications technology to the functioning of society is increasing (Krechmer, 2005), this source for FMA is becoming increasingly important. Telecom providers enjoying pole position on the learning and experience curve may be able to gain higher profits than they would be able to without this lead over the competition. Although – because of the typical network-characteristic of large sunk costs and low marginal costs per unit – there will usually be a drive to pre-empt as much demand as possible before competition enters the marketplace (asking for a penetration price strategy), to a certain extent the first mover position allows a price skimming strategy to recover its sunk costs quickly before competition steps in and lowers the market price.[4]

Secondly, the first-mover firm may be able to gain advantage by pre-empting rivals in the acquisition of scarce assets. A well-known example occurred in the UMTS-market, where telecommunication companies bought expensive UMTS-frequencies, which remained partly unused for a long time. By doing so, competitors were restrained from using those frequencies, which resulted in relative high consumer prices for usage of the scarce UMTS-frequencies that were actually exploited by the providers. Also, huge pre-emptive investments in ducts and dark fibre might be seen as an example of pre-emption, taking away incentives for new entrants to invest in this market segment.

Buyer switching costs in telecommunication markets are created by providers, similar to in a schoolbook example. Billing initial costs of connection, making customers pay upfront for provider specific gear (e.g. a modem), contractual binding of customers (e.g. by imposing a minimum contract duration or exit fees) are common commercial practices. With switching costs, late entrants must invest extra resources to attract customers away from the first-mover firm.

Lieberman and Montgomery (1987) also identified four sources of second-mover advantages (SMAs, also referred to as first-mover *dis*advantages): (1) the ability to 'free ride' on first-mover investments,

(2) the resolution of technological and market uncertainty, (3) technological discontinuities that provide 'gateways' for new entry, and (4) various types of 'incumbent inertia' that make it difficult for the incumbent to adapt to environmental change. These phenomena can reduce, or even completely negate, the net advantage that the incumbent derived from the mechanisms considered previously. In other words, by waiting a bit longer advantages appear like the possibility to free ride on the investments of pioneering telecom providers, for example by using infrastructure of an incumbent being obliged to service the new entrant on behalf of regulatory authorities. Late movers can gain an edge through resolution of market or technological uncertainty. In many new product markets, a dominant design often emerges over time and the demand situation becomes more clear. Pioneering telecom firms investing in optical fibre had to cope with a high degree of uncertainty about the emergence of sufficient bandwidth-intensive services to make it worthwhile realising fibre to the home. After the partial resolution of this uncertainty, second movers started investing in optical fibre infrastructure, facing less business risk. Also, as Gilbert and Birnbaum (1996) argued, after some time regulatory uncertainty usually diminishes because of the existence of more explicit policies of national regulatory authorities,[5] concrete rulings of and experience with regulators etc. Further on, we will consider this specific type of resolution of uncertainty next to the resolution of market and technology uncertainty. Information about shifting technology or customer needs supported new telecom entrants in their decision to stop investing in copper networks (DSL) but to shift towards optical fibre. Finally, there is the circumstance of incumbent inertia, which makes the incumbent less flexible at responding to changing customer needs or technology shifts, providing a gateway for other providers than the SMP-operator to play at leapfrog with the incumbent. High sunk costs laid down in the DSL network of an operator makes it unlikely that this operator will volunteer to invest in optical fibre networks; such cannibalism would only be justified under a concrete competitive threat.

To sum up, we may say that the examples above clearly show the relevance of advantages of innovation timing to telecom providers' strategy.

5.3 Network effects

In this paragraph some typical characteristics of telecommunication networks will be mentioned along with their influence on telecom providers' strategy.

One typical characteristic, which is common to most network sectors, is the combination of high sunk costs and low marginal costs, leading to extraordinary economies of scale. Sunk costs are investment costs incurred before a certain activity takes place, which cannot be recovered by the possible sale of the asset they produced. Highly specific investment (e.g. R&D, infrastructure) are usually sunk costs and represent barriers to exit. Marginal costs indicate by how much the total costs change because of modification in the production level by one unit (Piana, 2003). Servicing one more telecom subscriber on a telecom network usually does not substantially increase the provider's total costs. This network effect makes it essential for providers to realise a large installed base of customers in order to get a return on their network investments.

Secondly, telecommunication networks are subject to positive network externalities; consumers value a telecommunications system more as more users adopt it (Grubera and Verboven, 2001). This second network effect is not just relevant to any network. It can even be the other way around, for example when the network is used to distribute gas or energy, too many users using it at the same time might lead to shortages. Or, for example, with regard to public transport infrastructure, too many people in a train can lead to uncomfortable situations. However, if you are the only one with a phone on a telecom network, there is nobody to call, so the value of that telecommunication system for you is extremely low. The more people you can call, the more value your connection delivers for you. This means that individually switching to a new telecom provider is of no interest if there is no interoperability between the old and the new provider. If the old provider is an incumbent and new provider is the new entrant and the incumbent has no obligation to arrange interconnectivity with the new entrant's network, it is not likely that new entrants will be able to attract many customers from the incumbent. This is the second network effect. Further on we will refer to this phenomenon as 'community-effect'.

Given the combination of these two network effects, a logical first-mover strategy would thus be to gain the highest possible market share before competition enters the market (maximise economies of scale) and to prevent consumers using competing telecom networks to connect to its own subscribers, or at least to make it more expensive or more difficult for second-mover subscribers to connect to first-mover subscribers. By doing so, consumers experience advantages in subscribing to or staying with the provider with the largest customer base (community).

The first-mover strategy sketched in this section benefits from the mechanism that network effects enhance some FMAs and vice versa; market penetration and economies of scale contribute to the strength of the first mover's technological leadership while technological leadership might be useful for further market penetration and community-effects enhance buyer switching costs while buyer switching costs can lead to the creation of an even bigger community. Because of these enhancements, in an emerging telecommunication market, a first mover applying such a strategy is likely to become a provider with sustainable market power (a so-called SMP operator / incumbent). However, community-effects lose value under interconnection obligations; the value may even drop to zero when providers would additionally be prohibited from charging different prices for calls terminating on their own network compared to calls terminating on other networks.

So, to conclude, next to innovation timing advantages, network effects also matter in terms of strategic decision making by telecom providers – the effects of the two often boost one another.

5.4   Influence of regulation on the attractiveness of a new entrant's strategic options

In this section common regulatory practices in telecommunications will be confronted with innovation timing advantages and network effects to analyse the institutional influence of several regulatory practices on the attractiveness of new entrants' strategic options. To understand the influence of specific (changes in) regulatory provisions on telecom providers' strategies, it is essential to understand the sole influence first of innovation timing advantages and network effects as mentioned above, like reviewing a hypothetical situation where no telecom regulation would apply (yet) at all. In other words: looking at a telecommunication market in which an incumbent is present – which is likely because of the network effects enforcing FMAs as set out in section 5.3[6] – we are interested in the attractiveness of several new entrant strategies. Hereto, in Table 5.1 the FMAs (F1–F4), the SMAs (S1–S4 including S2′) and the network effects (N1–N2) are confronted with three new entrant strategies as mentioned in section 5.1: SGN, NGN and SBC.

In an unregulated situation with only one provider (the incumbent) offering telecommunication services using its own network, this incumbent is unlikely to be willing to share its network with new entrants in order to make SBC possible.[7] Without the SBC option, a new entrant may choose to enter into competition with the incumbent

Table 5.1   Attractiveness of new entrant strategies in an unregulated market

facing		attractiveness of new entrant strategies in an unregulated market		
		SGN	NGN	SBC
F1	technological leadership incumbent	–	+	–
F2	pre-emption of assets by incumbent	–	+	–
F3	buyer switching costs	–	–	–
F4	price skimming by incumbant	–	+	–
S1	free rider effects (inf. spillover)	+	–	–
S2	resolution of uncentainty (market/techn)	+	–	–
S2'	resolution of regulatory uncertainty	+	+	–
S3	shifts in tech/customer needs	–	+	–
S4	incumbant inertia	–	+	–
N1	expensive infra, low marginal costs	–	+	–
N2	community effect	–	–	–

by introducing a SGN or by investing in an NGN. The new entrant
might be able to develop a SGN under more efficient conditions than
the first mover had been able to in the past because of free-ride effects
and resolution of uncertainty (S1, S2 and S2'). However, because of
the incumbent's technological pole position, the assets it has already
pre-empted, customers unwilling to switch providers because of high
switching costs, the incumbent's ability to lower prices when competition evolves (F1–F4) and especially because of the FMA-enhancing
network effects (economies of scale and the community effect), these
second-mover advantages are not likely to overwhelm the first-mover
advantages. So, investing in a SGN appears to be unattractive. At the
same time, using changing technological and customer needs and making use of the incumbent's inertia (S3 and S4) to become a first mover
itself with regard to NGN technology and thereafter enjoy technological
leadership on the NGN in the long run, the possibility to pre-attempt
NGN-assets and to apply the prize-giving strategy in the future itself
(F1, F2 and F4) might be more interesting for the new entrant. If such
a leapfrog-enabling technology gateway is available, investing in an
NGN appears to be the most attractive strategy for the new entrant in
an unregulated telecommunications market.

Now we will consider the institutional influence of several common
regulatory practices on the attractiveness of the same new entrant's

strategic options. The scope and aim of this chapter only allows for the analysis of the influence of a limited number of illustrative examples of regulatory measures. There is no exhaustive pretention at all to consider any and all possible strategy-influencing and mutually interfering regulatory practices.

One important common regulatory practice in telecommunications is mandated access to the incumbent's network[8] to stimulate innovation of network services ('essential facilities doctrine'). Often, such an obligation is applied in combination with strict rules determining the maximum prices for the mandated access to prevent the incumbent from preventing competition on its network by setting sky-high access prices. Under these conditions, the attractiveness of the new entrant's strategies is influenced as illustrated in Table 5.2, where an arrow-up stands for positive influence, an arrow-down for negative influence and a zero for lack of significant influence in comparison with the unregulated situation.

If SBC is made possible and affordable by regulation, SBC becomes more attractive than an NGN because of F1 (the new entrant can enjoy the incumbent's technological leadership by using its network), F2 (previously pre-empted assets become indirectly available to the new entrant by using the incumbent's network), F4 (as a result of the price-regulating

Table 5.2 Institutional influence of cost-based mandated access

facing		attractiveness of new entrant strategies in an unregulated market			institutional influence of cost-based mandated access		
		SGN	NGN	SBC	SGN	NGN	SBC
F1	technological leadership incumbent	−	+	−	0	↓	↑
F2	pre-emption of assets by incumbent	−	+	−	0	↓	↑
F3	buyer switching costs	−	−	−	0	0	0
F4	price skimming by incumbant	−	+	−	0	↓	↑
S1	free rider effects (inf. spillover)	+	−	−	↓	0	↑
S2	resolution of uncentainty (market/techn)	+	−	−	0	0	0
S2′	resolution of regulatory uncertainty	+	+	−	0	0	0
S3	shifts in tech/customer needs	−	+	−	0	0	0
S4	incumbant inertia	−	+	−	0	↓	↑
N1	expensive infra, low marginal costs	−	+	−	0	↓	↑
N2	community effect	−	−	−	0	0	0

element of the mandated access obligations, the incumbent's price-skimming intentions won't affect the new entrant any more or the new entrant may even enjoy part of the consumer surplus by following the price-skimming strategy the incumbent applies to its own customers), S1 (no need to find out how to build an SGN if you can free ride on the incumbent's network), S4 (no need to get inert by investing in an NGN if the incumbent's network is available without high initial investments) and N1 (the SBC option takes away upfront investments in expensive infrastructure and maintenance, only 'pay for use' remains). All in all, under the circumstances mentioned above, cost-based mandated access leads to declining attractiveness in investing in NGNs in the first place and even takes away incentives to build an SGN (which were scarce already in comparison with the SMAs as shown above) as well. For example, there is no need to enjoy free-rider effects in building an SGN (S1) if such an SGN can be used on a SBC-basis. In other words: NGNs are more likely to emerge in unregulated telecommunication markets; cost-based mandated access seems to have a disastrous effect on the development of NGNs.

To make matters worse, the incumbent under cost-based mandated access obligations might become less willing to invest in its own network for two reasons:

1. Profits can no longer be totally pre-empted as part of the potential turnover floats to the service based competitor(s),[9]
2. The threat of emerging competition from an NGN-player declines as the NGN-scenario has become less attractive for a new entrant, so there is no need for innovations from a defensive point of view.

In the same way, we can assess the institutional influence of other common regulatory practices, like interconnectivity obligations, portability obligations, harmonising European regulatory power, universal service obligations for the SMP operator, facilitating rights of way etc. The confrontation matrix of these measures with innovation timing advantages and network effects are summarised in Table 5.3.[10]

The table shows us that interconnectivity obligations have a positive effect on the attractiveness of each scenario for the new entrant. Interconnectivity obligations break through the strong mutually strengthening effect of buyer switching costs (F3) and the community effect (N2); because of interconnectivity all subscribers to any provider form one big community resulting in community effects for all competing providers. For customers there is no need to stick with the provider

Table 5.3 Institutional influence of several common regulatory measures

facing	attractiveness of new entrant strategies in an unregulated market			institutional influence of interconnectivity obligations			institutional influence of portability obligations			institutional influence of harmonizing European regulatory power			institutional influence of universal service obligations for SMP operator			institutional influence of facilitating rights of way		
	SGN	NGN	SBC	SGN	NGN	SBC*	SGN	NGN	SBC*	SGN	NGN	SBC*	SGN	NGN	SBC*	SGN	NGN	SBC*
F1 technological leadership incumbent	−	+	−	0	0	0	0	0	0	0	0	0	0	0	0	0	0	0
F2 pre-emption of assets by incumbent	−	+	−	0	0	0	0	0	0	0	0	0	0	0	0	0	0	0
F3 buyer switching costs	−	−	−	↑	↑	0	↑	↑	0	0	0	0	0	0	0	0	0	0
F4 price skimming by incumbant	−	+	−	0	0	0	0	0	0	0	0	0	0	0	0	0	0	0
S1 free rider effects (inf. spillover)	+	−	−	0	0	0	0	0	0	0	0	0	0	0	0	0	0	0
S2 resolution of uncentainty (market/techn)	+	−	−	0	0	0	0	0	0	0	0	0	0	0	0	0	0	0
S2′ resolution of regulatory uncertainty	+	+	−	0	0	0	0	0	0	↑	↑	↑	0	0	0	0	0	0

(continued)

Table 5.3 Continued

facing	attractiveness of new entrant strategies in an unregulated market			institutional influence of interconnectivity obligations			institutional influence of portability obligations			institutional influence of harmonizing European regulatory power			institutional influence of universal service obligations for SMP operator			institutional influence of facilitating rights of way		
	SGN	NGN	SBC*	SGN	NGN	SBC*	SGN	NGN	SBC*	SGN	NGN	SBC*	SGN	NGN	SBC*	SGN	NGN	SBC*
S3 shifts in tech/ customer needs	−	+	−	0	0	0	0	0	0	0	0	0	0	0	0	0	0	0
S4 incumbant inertia	−	+	−	0	0	0	0	0	0	0	0	0	0	0	0	0	0	0
N1 expensive infra, low marginal costs	−	+	−	0	↑	↑	0	0	0	0	0	0	0	↓	↑	0	↑	0
N2 community effect	−	−	−	0	0	↑	0	0	0	0	0	0	0	0	0	0	0	0

SBC* = Service Based Competition provided that mandated access obligations are in place.

used by most of their friends, as it is possible to communicate with any subscriber from any provider under the same conditions.[11] However, providers with only a few subscribers benefit more from this effect than providers who already enjoy major (own) community-effects because of owning a large installed base already.

Portability obligations are regulatory measures aimed at establishing easy and inexpensive provider switches for customers. For example, shortening the maximal duration of telecom subscriptions, rules aimed at minimising the downtime a customer faces when switching between providers, obligations to make telecom offers easier to compare, enforcing standards (e.g. standardizing a high-speed modem for all providers of similar network technology) etc. all contribute to lowering buyer switching costs (F3), which is positive for new entrants irrespective of the chosen strategy. The same applies to reducing regulatory uncertainty, for example by harmonising the interpretation and application of European directives by national regulatory authorities. The current development to further harmonise European regulatory power (replacing the European Regulators Group (ERG) with the Body of European Regulators for Electronic Communications (BEREC), which was aimed at improving the consistency of implementation of the EU regulatory framework) might contribute to this.

SMP operators are – more often than providers with no sustainable market power – encumbered with universal service obligations. In general, companies are not very keen on these kinds of obligations as they tend to limit entrepreneurial freedom and they usually do not substantially contribute to the firm's profitability. From this perspective, SBC is more attractive than becoming an SMP operator or an NGN.

To conclude, facilitating rights of way positively influences the SGN and NGN strategies of a new entrant because of savings on infrastructure building costs. For implementing an SBC-strategy, rights of way are less relevant provided that the necessary network connections have already been established.

5.5   Conclusions, reflection and suggestions for further research

First of all, this chapter shows that the structured application of innovation timing theory combined with the concept of network effects in the telecommunications industry and related to new entrance strategies provides a better understanding of the variegated influence of several regulatory measures. Also, when reassessing regulatory regimes, the framework can support regulatory decisions by providing more structural insights into the relation between regulation and new entrance strategies.

The instigated framework, for example, contributes to a better understanding of the current debate about next generation access networks; traditionally, European regulatory policy strongly promoted mandated access obligations. Recently, Viviane Reding (former member of the European Commission responsible for the Information Society and Media) stated:

> Some argue that we need a regulatory forbearance to encourage investment in new network infrastructure. But in my view there is no evidence to show that 'regulatory holidays' would generate more investment than competition. We have to be clear on this point: regulation should promote competition, and should not favour monopolies. Investment in new and competing infrastructure will accelerate the day when transitional access obligations can be further relaxed – or even removed. Richards e.a., (2006).

The argumentation in this statement seems to contradict the conclusion above that mandated access obligations can seriously hamper the emergence of NGNs. In this respect, we would like to draw attention to a development in the United States where the Federal Communications Commission (FCC) recently ruled that telecom providers can build fibre-optic networks into certain (non-commercial) buildings without sharing their infrastructure with rivals:

> As part of the 1996 Telecommunications Act, the Baby Bells are required to lease or 'unbundle' access to their copper network wires, which enable telephone and dial-up Internet services. But last year, the FCC ruled that new broadband networks built with fibre optics to serve single-family homes would not be subject to the same sharing requirements imposed on the older copper wire infrastructure. In March, the US Court of Appeals for the District of Columbia upheld the FCC's ruling. This week's ruling extends the reach of the original by addressing multitenant buildings. Michael Powell, chairman of the FCC, [...] said that by clarifying the rules, big phone companies now have more incentive to deploy these fiber networks, particularly in urban areas. (Reardon, 2004).

Also within Europe, more and more voices are being raised in favour of limiting the historically uncompromising mandated access obligations. The draft recommendation of the European Commission to national telecom regulators (see footnote no. 6) identifies different kinds of business arrangements that operators can pursue to share the risk of developing new networks, while still abiding by EU competition law. The arrangements include joint-investment by different companies, long-term access contracts between larger and smaller operators, and 'volume discounts' for operators that want to rent a relatively large chunk of network capacity (Brunsden, 2009).[12]

On the basis of the framework instigated in this chapter, we may contribute to this European discussion by stating that relaxing mandated access obligations, as in the recommendation or even further, appears to influence the emergence of NGNs positively.[13] Furthermore we may state that interconnectivity and portability obligations not only guarantee more regulatory certainty but also have positive effects on all new entrance strategies.

Apparently, interconnectivity and portability obligations also contribute to lower prices and more freedom of choice for consumers. So, this 'public value' goes hand in hand with stimulating more innovative strategies for new entrants. However, relaxing mandated access obligations might lead to higher prices and less freedom of choice for consumers as a result of the lower attractiveness of the SBC option. Here we end up at the very fundamental challenge of optimal balancing public values in the short term (lower prices and more freedom of choice for consumers) and public values in the longer run (enjoying the benefits of more innovative networks). At this point further research on the area of 'smart rules' is of vital importance.

Although constructing a framework as instigated in this chapter appeared to be a fruitful exercise, the framework needs to be refined in a more sophisticated way. Currently, hardly any attention has been paid to mutually interfering measures, to the protection by measures under consideration of other public values than innovation, to the influence of existing and potential other regulatory measures (e.g. functional separation of networks and services) etc. Also, the framework in its current appearance only applies to markets where only one incumbent (network) is available. Furthermore, the current framework is based on and limited to qualitative insights though more quantitative support would be valuable. Finally, cross-effects of non-institutional influences should be considered as well. All in all, some refinement still has to be done, but the effort promises scientifically and economically relevant decision-supporting knowledge about the relation between regulation and innovation.

Notes

An earlier version of this chapter was presented by the author as a conference paper in Chennai (India) on 11 December 2009 at the 'Developing 21st Century Infrastructure Networks conference, sponsored by Next Generation Infrastructures Foundation, Center for Study of Science, Technology and Policy (CSTEP), SSN College of Engineering, and the IEEE Systems, Man & Cybernetics Society.

1. NGNs are commonly built around the Internet Protocol, and therefore the term 'all-IP' is also sometimes used to describe the transformation towards NGN (2009).
2. Given this assumption concerning the initial market situation, this paper might be of less relevance for regions with different initial conditions.
3. Strictly speaking, *early* mover advantages might be a better label than *first* mover advantages; apart from technological leadership, FMAs can be available to more 'early movers' instead of exclusively to the first mover (Gilbert e.a., 1996), just like SMAs can be available to several 'late' movers. Sometimes, SMAs are not at all available to the second mover as such but evolve later in time and can only just be enjoyed by actor(s) introducing their product or service even later in time than the second mover did. So, for a second mover, it is possible to innovate 'too early' resulting in neither FMAs nor SMAs. It is also possible to innovate 'too late' to gain any SMAs (Gilbert e.a., 1996). In this chapter, by first mover we also intent to address other early movers gaining 'first mover advantages' and by second mover we mean any mover gaining second mover advantages.
4. Given an inelastic demand curve for the specific telecom service.
5. A concrete example is the current attempt of the European Commission to come up with a recommendation for National Regulatory Authorities in telecommunications about how to deal in the near future with regulated access to NGNs (better known as Next Generation Access Networks).
6. Historically, many telecommunication networks emerged as a public service and recently, most of those networks have been privatised. The private provider exploiting such a network often possesses an incumbent position anyway, also because of this historical background.
7. However, deviating business strategies are conceivable, like incumbents offering network-services (voluntary and of course profit-based in the longer run instead of mandated and cost-based) to new entrants, in order to shorten the ROI of the physical network and to prevent new entrants from building a competing network. Like this, SBC might become an option for a new entrant. Nevertheless, as long as network-based competition remains absent, the monopolistic power of the incumbent to determine tariffs and conditions might make this option unattractive. A different situation might occur if a network owner decides *not* to offer services to end users itself but to leave this up to separate service providers ('functional separation'). As this situation deviates from the initial market conditions as described in this chapter as point of departure for the analysis, this scenario won't be discussed further here.

8. For example, mandating access is considered as a means of increasing competition in Directive 2002/19/EC ('Access Directive', 2002).
9. It goes without saying that this might be totally different if several competing networks are open to SBC, in which case a network access market arises in response whereto network operators might improve their networks to offer a better service to service based competitors (see also Bijlsma and van Dijk, 2007). Like this, they try to attract as many end users to their network as possible and secure at least a little indirect turnover from those users.
10. SBC* stands for Service Based Competition provided that cost-based mandated access obligations are in place. When assessing other regulatory measures, it makes sense to assume that cost-based mandated access obligations are already in place, otherwise SBC would be no option anyhow.
11. Whether or not these conditions are the really the same, depends on the commercial freedom that regulators leave at the providers to set different termination fees for cross-provider connections than for establishing connections within their own network.
12. According to Innocenzo Genna, chairman of The European Competitive Telecommunications Association (ECTA), which represents many smaller telecoms operators, the arrangements could 'allow dominant firms to escape regulation, limiting choice in TV and broadband services' and that the Commission 'appears to have compromised its strong stand against regulatory holidays in the telecoms sector and appears to be condoning collusion'. According to the European Telecommunications Network Operators' Association (ETNO), which represents large operators such as Deutsche Telekom and Telefónica, has criticised the Commission for not offering enough regulatory flexibility ... Michael Bartholomew, ETNO's director, said that 'a more innovative and targeted regulatory approach, which encourages all players to invest and share risk' is needed (Brunsden, 2009). These opposite opinions perfectly fit in the framework which clearly distinguishes the incumbent positions from the new entrants'.
13. However, once the NGN emerged in a certain region, further development of services on that network (service based competition) might be stimulated by imposing mandated access obligations in a next phase. Granting a 'regulatory holiday' might thus be effective for a NGN to come into being, provided that such a holiday is granted in advance (in order to secure regulatory certainty) for a period long enough to get satisfying return on network-investments.

References

Bijlsma, M. and M. van Dijk, 'Nieuwe generatie netwerken, nieuwe generatie regulering?', *CPB document 145*, May 2007.

Brunsden, J., 'Telecoms firms hit out at network guidelines'. Available at: http://www.europeanvoice.com/ article/2009/06/telecoms-firms-hit-out-at-network-guidelines/65181.aspx, 12 June 2009.

Directive 2002/19/EC of the European Parliament and of the Council of 7 March 2002 on access to, and interconnection of, electronic communications networks and associated facilities (Access Directive), consideration 19.

Gilbert, J. T. and P. H. Birnbaum-More, 'Innovation timing advantages: from economic theory to strategic application', *Journal of Engineering and Technology Management*, vol. 12 (1996), pp. 245–66.

Grubera, H. and F. Verboven, 'The evolution of markets under entry and standards regulation: the case of global mobile telecommunications', *International Journal of Industrial Organization*, vol. 19 (2001), pp. 1189–212.

Irwin, A. and P. Vergragt, 'Re-thinking the relationship between environmental regulation and industrial innovation: The social negotiation of technical change', *Technology Analysis & Strategic Management*, vol. 1/1 (1989), p. 58.

Krechmer, K., 'Communications Standards and Patent Rights: Conflict or Coordination?', presented at The Economics of the Software and Internet Industries conference in Toulouse, January 2005.

Lieberman, M. B. and D. B. Montgomery, *First-Mover Advantages*, Research paper 969 (1987).

Lundvall, B-Å. I., *National Systems of Innovation: Towards a Theory of Innovation and Interactive Learning* (London and New York: Pinter Publishers, 1992), pp. 139–40.

NGN, online available at http://en.wikipedia.org/wiki/Next_Generation_ Networking, 25 June 2009.

Piana, V., 'Costs', Economics web institute, 2003, online available at http://www. economics-webinstitute.org/glossary/costs.htm#cost, 29 June 2009.

Reardon, M., 'Baby Bells win another FCC victory', *CNET News*, (2004).

Richards, E., R. Foster, and T. Kiedrowski, 'Communications the next decade: A collection of essays prepared for the UK Office of Communications', (2006), p. 6.

Standardization Sector of the International Telecommunication Union, *General overview of NGN*, ITU-T Recommendation Y. 2001, 2004.

6

Deutsche Telekom and Pacific Bell v. LinkLine: Does Competition Law have a Place in Regulated Industries?

Martin W. Holterman

6.1 Introduction

In the last two years, both the US Supreme Court and the European Court of Justice have been faced with the problem of what to do when competition law and telecommunications regulations collide. In *Deutsche Telekom v. Commission*,[1] the European Court of Justice (hereafter: ECJ) agreed that Deutsche Telekom could be guilty of abusing its dominant position on the market for DSL Internet connections by carrying out a so-called price squeeze, whereby a supplier that competes with its own customers seeks to increase its market share by reducing the retail/wholesale margin, that is, the difference between the price charged to its wholesale customers and the price it charges its own retail customers for the finished product. Interestingly, the appellant's objection that its pricing system was, in any event, approved by the German telecom authorities was deemed irrelevant.[2]

On the other hand, when the Supreme Court of the United States was faced with the same problem in *Pacific Bell v. LinkLine*,[3] it ended up finding for the monopolist. Even though the American Telecommunications Act explicitly preserves the possibility of bringing an antitrust suit against a telecom company,[4] the court held that in this case the Sherman Act had not been violated.

In this chapter, the main focus is not on the legal arguments underlying each court's approach, which are generally beyond reproach given each jurisdiction's respective statute and case law, but on the on the pragmatic question of whether it would be appropriate for the competition authorities to interfere in the DSL industry in this way. After all, the law already provides for a regulator, so why would it be necessary or even appropriate to add a second? Depending on how their relative powers are arranged, the competition authority may well get in the way of the more knowledgeable regulator.

Even if we presume that the competition authority has a similar level of understanding of the market, it is still likely that it will have different policy goals than the regulator. We might speculate, for example, that the regulator will care more about consumer protection or that the competition authority will focus disproportionately on static instead of dynamic efficiency. While it is not intended that the present chapter will carry out any kind of empirical survey of the attitudes of various government agencies, it is surely useful to explore what the consequences would be of a difference in perspective.

This problem of run-ins between government agencies was quite rare until recently, which explains the almost total absence of literature, at least the normative kind.[5] However, with the ever increasing popularity of liberalisation and privatisation, as well as the increasing activism of many competition authorities, we will likely see much more of it in the future. For this reason, it is important that we should come to terms with this situation, not only from a legal perspective, but also from the perspective of the social sciences.[6] In this chapter, I will examine in particular what the consequences for innovation are of a run-in between regulation and competition law.

In what follows, I will begin by briefly discussing the *Deutsche Telekom* and *Pacific Bell* cases as examples of competition law 'intruding' in a regulated industry. Subsequently, in sections 6.3 and 6.4, I will develop a simple model of regulation and innovation, based on certain simple assumptions about how each agency might behave. In section 6.5 I will then examine the 2002 EU Telecommunications package,[7] as well as the choices made by the various actors in the *Deutsche Telekom* case in order to discover what kind of role the EU legislator intended for the competition authorities, and whether the European Commission played along. Ideally, of course, my model of section 6.4 should be consistent with their analysis. Section 6.6 concludes.

6.2   The cases

Before continuing to the modelling part of this chapter, it is good to examine the *Deutsche Telekom* and *Pacific Bell v. LinkLine* cases in a bit more detail. That way, we can later shift back and forth between the abstract and the concrete more easily, which will be useful in order to maintain at least some measure of realism and practicality.

Pacific Bell v. LinkLine (2009)

The plaintiff already passed the first hurdle in *Pacific Bell* before the hearing of the case even began. In the 2004 case of *Verizon v. Trinko*,[8] the court had already ruled that the mere fact that there is such a thing as a specialised Telecommunications Act[9] does not mean that antitrust suits are foreclosed. Ordinarily, the application of the doctrine of implied immunity[10] to the telecommunications market would be a very real possibility, but Congress explicitly preserved the applicability of the antitrust laws to this industry.[11] Whether it was wise of Congress to accept 'the real possibility of judgements conflicting with the agency's regulatory scheme that might be voiced by courts exercising jurisdiction under the antitrust laws' is, of course, the topic of this chapter.[12]

Having thus ruled out the easy *lex specialis* solution, the next question is which theory of antitrust liability the court should adopt. Of course, the Sherman Act only says that 'Every person who shall monopolise [...] any part of the trade or commerce among the several States [...] shall be deemed guilty of a felony',[13] without specifying what kinds of behaviour constitute monopolisation, but in practice it is a lot easier to win an antitrust case if you can frame it in terms of a well-known antitrust theory, be it vertical price maintenance, unlawful tying or something else again. Each of these monopolisation theories has a line of case law to back it up, meaning that the courts know exactly what to look for. A plaintiff looking for the court to recognise a new kind of monopolisation is always going to face an uphill battle.

In *Pacific Bell v. LinkLine*, what the plaintiff would have liked is for the court to treat margin squeezing as a distinct antitrust theory, which is what the lower courts did.[14] Unfortunately, none of the justices were prepared to do so. Instead, they framed a margin squeeze as a combination of refusal to deal in the wholesale market and predatory pricing in the retail market. In the former market, the price was too high, and in the latter, it was too low. The question was whether LinkLine could use the antitrust laws to make sure the prices were set just right. For the wholesale market, that meant all but relitigating *Trinko*, and it was only to be expected that it failed; the court found that Pacific Bell did not have an obligation under the antitrust laws to sell to anybody in the first place, meaning that it could sell to whomever it liked at whatever price it liked, as far as the Sherman Act was concerned. The fact that there was an obligation to sell under the Telecommunications Act was deemed irrelevant, as it had been in *Trinko*. As for the predatory pricing theory, the *Brooke Group* precedent dictates that it is insufficient that the monopolist charges a price that is below an appropriate measure of costs, which is the only requirement in Europe.[15] Instead, the plaintiff also has to demonstrate that 'there is "a dangerous probability" that the defendant will be able to recoup its "investment" in below-cost prices'.[16] This is a hurdle only few plaintiffs are able to pass, and LinkLine, too, failed to impress the court.

In this way, the result is that regulation and antitrust law are kept apart quite neatly, not because of some principled decision to that effect, but simply because the Supreme Court was worried about 'chill[ing] the very conduct the antitrust laws are designed to protect'[17] in a regulated market as much as it would be in any other market. Even though another outcome is not categorically foreclosed, it is unlikely that the Federal Communications Commission will have to tolerate competition from the antitrust authorities very often.

Deutsche Telekom v. Commission (2010)

On the other side of the ocean, the ECJ tackled the same problem in *Deutsche Telekom v. Commission*,[18] and more recently in *Konkurrensverket v. TeliaSonera*.[19] In these cases, too, there was a regulated dominant incumbent, heir to a previous monopoly, accused of carrying out a margin squeeze in violation of the competition laws.

Just like the plaintiffs in the American case, the European Commission had a very straightforward argument at its disposal to avoid the *lex specialis* rule: the European competition rules are based on the Treaties themselves, specifically art. 102 TFEU. Telecom regulation, on the other hand, is found only in secondary Community law and in national implementing legislation. As a result, the latter can never trump the former, simply as a matter of hierarchy of rules.[20] The only way this could be different is if the national regulatory framework had left Deutsche Telekom entirely without freedom of manoeuvre,[21] in which case the appropriate remedy would have been for the Commission to commence infringement proceedings against Germany.

The Commission did have one advantage: there was some precedent in European Community Law for a margin squeeze theory of antitrust liability, albeit not much. In 1988 the Commission had used that same theory to impose a fine of 3 million ECU on British Sugar, a decision that had gone uncontested,[22] and in 2000 the Court of First Instance signed off on it in the context of a rather curious anti-dumping/abuse of dominance case.[23] Despite this paucity of jurisprudence, Deutsche Telekom barely even attempted to argue against margin squeezing as a distinct theory of anticompetitive behaviour.[24] After all, there was enough case law on abusive pricing practices generally to make that a non-starter.[25] Given such precedents, there was little chance that the Community courts would be as worried as their American counterparties about discouraging companies from setting their prices low. When all was said and done, Deutsche Telekom lost on both the facts and on the law, and was ordered to pay a fine of € 12,6 million.

Then there is the case of *TeliaSonera*, which came to the court in the form of a prejudicial question.[26] The benefit of that format is that it involves giving the national court the benefit of the doubt before it rules, instead of applying a degree of deference to the Commission after it has already decided. This is done by explicitly stating the general rules that underlie the court's judgements, as well as by stating the outer boundaries of what would qualify a reasonable application of Community law by the national court. In this way we learn that it is possible for a margin squeeze to have an 'objective justification', which would take away its abusive nature.[27] Also, it might be the case, for example, that the margin squeeze does not actually reduce competition, because the wholesale product is not truly indispensible.[28]

The final issue discussed by the court in *TeliaSonera* is of particular relevance here: '[what is] the relevance [...] of the fact that the markets concerned are growing rapidly and involve new technology which requires high levels of investment'?[29] The court held that this was not relevant. To see why, we will need a quick primer on regulation, economics, and innovation.

6.3   Regulation, economics and innovation

The most widely used economic model of innovation is the model of Aghion and Howitt (1992),[30] which offers us an extended version of Joseph Schumpeter's creative destruction.[31] They argue that the connection between monopolisation and innovation consists of two opposing forces: on the one hand, the more companies that are allowed to be creative in the ways they obtain a dominant position and in the ways they use it, the greater their monopoly profits will be. The greater the monopoly profit associated with a successful innovation, the more companies will invest in R&D, and the more innovations will ensue. On the other hand, once an innovation has arrived, and produced a position of market power for its owner, this new monopolist has a vested interest in slowing down the arrival rate for future innovations.[32] If you are Netflix and you just put Blockbuster out of business by developing a state of the art distribution system for DVDs, the last thing you want is for people to switch to streaming video.[33]

As such, creative destruction means a battle between a powerful incumbent and infinitely many small innovators, between a bear and a swarm of bees, only with less cooperation among the bees. For this reason, the regulation of high-innovation industries is fiendishly difficult. Every time the regulator pushes back against the monopolist, it simultaneously reduces the monopolist's ability to stifle innovation and the incentive of innovators to innovate. After all, the latter is almost by definition impossible for the regulator to observe, while the monopolist is going to do its best to hide its innovation-stifling activities from the authorities.[34] And all of this is before we even start talking about regulatory capture.[35]

What the ECJ did in *TeliaSonera* was to focus exclusively on the second half of the story: the monopolist who stifles innovation. Starting from the observation that the letter of the law does not make any distinction between high-innovation or 'immature' markets on the one hand, and 'mature', low-innovation markets on the other hand, the court considered whether there was any reason to nevertheless make such a distinction. Starting from a point of view that – as usual in ECJ jurisprudence – comes dangerously close to seeing competition as a good thing for its own sake,[36] the court went on to emphasise the importance of making sure that one company would not be able to permanently take over a market by shoving out its competitors while the market was still 'immature', as well as the importance of watching out for monopoly leveraging, the practice of using monopoly power in one market to take over another.[37] After all, unlike the market for high speed internet access, the market for local loop access – the last part of the telecommunications network that connects the subscriber's house to the broader network – is 'in no way new or emerging, [though] its competitive structure is [...] still highly influenced by the former monopolistic structure'.[38] In such circumstances, the court argued, monopoly leveraging would be particularly harmful, meaning that there is no reason for the competition authorities to ignore 'immature' markets such as the market for DSL services.

This is all well and good, but it does not explain why – as a policy matter – the enforcement of competition laws in regulated industries cannot be assigned by the lawmaker to the regulator. The first and most obvious answer is that it can be. In the Netherlands, for example, the Competition Authority NMa is also responsible for regulating the public transport and energy industries. However, realistically that is only half the story. What actually happened is that the previously independent transport and energy regulators were merged with the NMa while continuing to exist as separate 'chambers'. In other words, as far as the day-to-day goings on are concerned, the competition authority and the regulator are as separate as ever.[39]

Apparently, and this goes for private as well as public organisations, there is a limit to the extent to which you can assign more than one task to one and the same group of people. This is not simply a matter of classical agency problems, which spell doom for any principal who assigns his agent two or more tasks that are to some extent mutually exclusive without telling him how to deal with any conflicts.[40] Instead, the problem is that regulators and competition authorities have a fundamentally different way of looking at the world, and trying to force one organisation to do both can only lead to abject failure.

A useful private sector analogy is the difference between preparing the accounts of a company for reporting purposes and preparing them for tax reasons. Not only are the accounting rules slightly different in most jurisdictions, so is the entire goal of the exercise; do we want profit to be high or low? To avoid a tragic case of schizophrenia, it is usually better to hire separate tax attorneys and accountants.

In what follows we will not explore this issue empirically, that is a task for another day. Instead, we will take this difference in perspective between the competition authority and the regulator as a starting point in order to see whether interference by the competition authority in a regulated industry is necessarily harmful to innovation, and to social welfare generally. Obviously, if the regulator has perfect knowledge and perfect incentives, involving the competition authority can only make things worse, but if their understanding of the industry is imperfect in different ways, and if their goals are similarly different and imperfect, adding a bit of 'crude' competition law to the mix may well make things better. In the next section, I will construct a model of two agencies hovering over one market in order to see when that might be the case.

6.4   Two helicopters and one market[41]

The first stylised fact that we will adopt is that the Regulator (hereafter: R) looks further to the future than the Competition Authority (hereafter: CA). In a model with three periods, R looks at the total picture, while CA only looks at the next period. This is tantamount to saying that R is better at promoting innovation than CA is. However, they are far from perfect, because like all government agencies they struggle to acquire the knowledge necessary to enforce the rules. Neither can say with perfect certainty whether the actions of a given company are beneficial to social welfare – either today or in the future – or not.[42]

As if this was not enough, we will also burden the agencies with a second problem: they have to act consistently within the rule of law. That means that they may not discriminate between companies that are – by all appearances – similarly situated, and that they must work from clearly defined laws. Note, however, that this does not mean that they must work from the same laws.

These two agencies play a sequential game with an innovative company trying to decide how much to invest in research and development. At $t=0$, the company observes the actions of the agencies, and decides to invest or not. At $t=1$, if it has invested, it makes the discovery, which gets implemented with some measure of anti-competitive behaviour and starts to pay off by $t=2$.[43] However, at $t=1$ CA gets another chance to intervene or not. F already decided what to do at $t=0$, and is required to stay the course, but at that time CA could not commit beyond $t=1$, given its limited ability to look ahead.

Given this setup, there are three possible outcomes:

A. The agencies intervene too heavy-handedly at $t=0$, the company declines to invest, and no innovation occurs.
B. The company invests, and CA intervenes.
C. The company invests, and neither agency intervenes.

The really interesting outcome here is B), the outcome where the government (and society) can have its cake and eat it, too. In this outcome, society – either through a fine or through lower prices on the market – siphons away the monopoly profits that would otherwise be the prize for a successful innovator. It is important to realise that this is not simply a net transfer of wealth. It is certainly that, but some of this transfer also comes in the form of reducing the innovator's dominance in the market, meaning that consumers get lower prices (an improvement in static efficiency and a reduction in the deadweight loss) and that future innovators get lower barriers to entry (an improvement in dynamic efficiency). This means that B) is the outcome that society prefers above the other two, at least as long as we stay within our three-period framework.

The company, on the other hand, prefers option B) least of the three. After CA is done, the company is worse off than if it had not invested in R&D in the first place. So if CA intervention is a certainty, the only rational thing for the company to do is to refrain from innovating. By no stretch of the imagination is B) a sub-game perfect (Nash) equilibrium.

This puts the ball back in the government's court. If it always prefers C) over A), it can achieve that outcome by enacting a law that keeps CA away from the industry. Similarly, if it prefers A), the legislator can encourage CA to do its thing. Unfortunately, this conclusion amounts to putting the cart before the horse: it requires that the legislator should fashion a one-size-fits-all solution to all innovations in the market at some time before $t=0$, when not even R knows what kinds of innovations are likely to occur.

Fortunately, there is another possibility: the legislator could make CA's choice uncertain. In a perfect world, the only way to do that is to tell CA to toss a coin before intervening, to deploy what is known in game theory as a 'mixed strategy'.[44] In the real world, on the other hand, the legislator need not do anything quite so strange as to order CA to refrain from intervening even when it would otherwise be inclined to do so. Instead, the legislator can simply rely on CA's inherent shortcomings. After all, we already agreed that both R and CA can only observe the company's actions with great imprecision. To begin with, we could imagine that this means that the CA has a p% probability of correctly identifying C) if and when it occurs.

Given that assumption, the choice of the company is simple. Letting $\pi(s)$ represent the company's profit in each of the three possible outcomes, including its expenditure on R&D, if any, the company will innovate as long as $(1-p)\ \pi(C) + p\ \pi(B) > \pi(A)$. This inequality reflects our assumption that CA will never be confused when the company declines to innovate. When outcome I, the status quo, is chosen, CA will always observe this correctly. As a result, when the company chooses outcome I, it is certain that it will get $\pi(A)$. Alternatively, it can innovate and gamble that CA will fail to recognise its anti-competitive behaviour, a gamble that will pay off with probability $(1-p)$.

It is easy to see that this reform induces the company to innovate more and to engage in more anti-competitive behaviour. However, the result is far from perfect. Assuming, for simplicity, that the company's profit in outcome B) is zero, we see that innovation will only occur if it produces an increase in profit of at least $\frac{p}{1-p}$%. For the coin toss example of $p=0.5$, this means that the only innovations that are undertaken are those that at least double the company's profits. While that is better than no innovation at all, one imagines the legislator would prefer it if smaller innovations also got to see the light of day.

Fortunately, the imperfection of CA will probably work differently in practice. Instead of having a constant error rate, one would expect them to have a higher error rate for smaller innovations than for large ones. In other words, one would expect p to be increasing in $\Delta\pi$, the difference in profit between C) and A). In that case, we can imagine a CA where $p(0)=0$, while for an arbitrarily large $\Delta\pi\,p$ is still smaller than 1; no matter how big the innovation, there is still a chance that CA will miss it. For intermediate-sized innovations, the key question is whether $p(\Delta\pi)$ is larger or smaller than the company's cut-off probability of $\bar{p} = \dfrac{\Delta\pi}{\pi(C)}$.[45] If the actual probability that CA will recognise that an innovation occurs stays at or below this level for all $\Delta\pi$, the company will always innovate.

The question is whether this is a good thing. Given that we started from the perspective of Aghion and Howitt (1992),[46] the answer is obviously no. Giving the company more freedom to do as it likes will cause it to innovate more, but it will also allow it to foreclose the innovations of others more effectively. For this reason, the legislator would want CA to be at least vaguely competent, so that the company will engage in innovation only part of the time.

So where does this leave the R, which already has to make up its mind at $t=0$? If nothing else, this means that R can pre-empt the decision of CA, at least in one direction; it can cause outcome A) to occur regardless of what the other agency thinks, by unilaterally turning the screws on the company. But would it actually do that?

For simplicity, we can start by assuming that R sees only the company's future profits π, just like CA did. Of course CA had no reason to care about anything other than the company's profit, since its sole goal is to use a combination of fines and structural measures to reduce the profit of a successful innovator to zero. R, on the other hand, would like to have information about the deadweight loss caused by the company's possible future monopoly power, as well as other measures of static and dynamic efficiency. However, to begin with we can assume that R has only the information about the company's possible future profits that the company itself also has.

Given that assumption, it is easy to see that R will never intervene. As a government agency, it does not care whether CA intervenes at $t=1$, since such intervention will never make society worse off. R will simply base its intervention decision on the comparison between $\pi(A)$ and $\pi(C)$, being the best proxies that it has at its disposal. But whenever $\pi(C)$ is smaller than $\pi(A)$, there is no need to intervene because the company would not innovate anyway. For this reason, if R has no better information than π, it will never intervene.

Fortunately, there is no reason to be quite so pessimistic. After all, the advantage that R has over CA, that is, that it can see further into the future, also has a side effect today: unlike CA, R can also see the profits of companies that are not yet dominant, and perhaps never will be. That means that, at the very least, R can observe the consequences of its interventions for the future total profitability of the industry. And even though R, like all government agencies, has to treat all companies equally, this does not mean that it has to treat all innovations equally. It is not suggested that the companies are competing to introduce the same innovation. On the contrary, the notion of Schumpeterian competition suggests that they will each have different innovations that they are looking to foist upon the consumer. R's rule making has the effect of ruling out some of these innovations, even if that is not R's intention. When R (and CA) are finished, the remaining innovations will fail or succeed depending on the vagaries of the market.

And that is the extent of the interplay between the regulator and the competition authority. R chooses which innovations to foreclose, knowing that any innovations that it might allow to proceed unmolested may or may not be accosted by CA. The company, in turn, has a different perspective than either R or CA. Whereas R cares only about the probability that CA will intervene because intervention will generally include structural measures that affect the position of the company's competitors, the company also cares about CA intervention to the extent that it involves simple fines. CA needs R because only the latter sees the bigger picture, and R needs CA because R is not allowed to intervene at $t=1$ because the law requires that legitimate expectations should be protected.

6.5 Back to the law

The previous section outlined a general argument that might explain why the legislator would let an admittedly inferior agency near a market that is already regulated. What is left to do is to consider to what extent this argument resonates with the actual practice of public administration and law, and particularly the practice as we have discussed it in section 6.2. As a policy matter, what kind of role did the European lawmakers intend for the competition authorities when they enacted the relevant sectoral regulation, particularly the Telecommunications Package of 2002?[47]

Looking at this legislation, it seems that the European legislator took a position more akin to our regulator than to the lawmaker as described by the model. Rather than setting up a regulatory model that would function no matter which way the technology developed, the EU institutions wrote laws that are very much rooted in the technology of today. They looked at today's telecommunications market and decided on a general framework of regulation, including many detailed rules. At the same time, they decided to make regulation only an instrument of last resort, with preference being given to treating the telecommunications industry like any other, subject to the ordinary competition rules.[48]

In the Explanatory Memorandum of the proposal for the Access Directive we read, for example: 'The Directive builds on the premise that competition rules will be the prime vehicle for regulating the electronic communications market once the market becomes effectively competitive'.[49] Similarly, in Recital 13 of that directive: 'The aim is to reduce *ex ante* sector specific rules progressively as competition in the market develops'. [50] Only during the transitional phase, while the old monopolists still controlled the market, did the legislator envisage any serious role for the regulator, whose job it is to ensure 'that bottlenecks in the market do not constrain the emergence and growth of innovative services that benefit users and consumers'.[51]

From this we could argue that in its *Deutsche Telekom* decision[52] the Commission simply played the part assigned to it. We could speculate that it did not ignore the possibility that the German regulator RegTP had given Deutsche Telekom some freedom to engage in margin squeezing on the theory that this was necessary to induce it – and other telecom companies – to innovate, but rather that it used its pre-existing power under the Treaties in order to do its part to minimise the ability of the incumbent dominant company to engage in foreclosing behaviour, as envisaged by the legislator/regulator at the European level.

Of course, there are also other possible reasons why the Commission did not defer to RegTP. One possibility is that the Commission thought its intervention was necessary because the regulator lacked the legal tools to reign in the telecom giant. That seems to be belied, however, by the fact that the history of German telecom regulation seems to be full of price caps and other measures that would have been necessary to avoid a margin squeeze.[53] The structural measures that are the sole province of the competition authority were not deemed necessary here by either agency.[54]

On the other hand, it is also possible that the Commission felt that RegTP simply got it wrong. If that is the case, they need not necessarily have said so explicitly, much less brought an infringement case against

Germany. After all, the relationship between the European Commission and the authorities in the Member States is often one of delicate diplomacy. Saying that a national regulatory authority is incompetent or in the pocket of industry will hardly help keep the mood high.

In any event, it is clear that the appearance of a legislator/regulator is not fundamentally problematic. Although the unique institutional setup of the European Union complicates matters somewhat, in general there is no reason why the legislature, in designing the institutional framework, cannot usurp some of the regulator's prerogatives. There is just one important condition: the legislative cycle has to be fast enough to keep up with the rate of innovation. It is for this reason that rule making is often delegated to government agencies in the first place: generally, they are able to amend their rules more quickly than the legislature would be able to.[55] Time will tell whether the European legislator's confidence is justified in this instance.

6.6 Conclusion

While the original question was whether there was a place for competition law in regulated industries, it turns out the better question may be whether there is a place for regulation in regulated industries. As we have seen, in the United States the courts keep antitrust plaintiffs on a short leash in network industries exactly because they are worried about harming incentives for innovation.[56] Regardless of the merits of that view, which arguably gives too little weight to the power of monopolists to foreclose future innovation, it at least has the effect of making successful antitrust suits in network industries a relatively rare phenomenon.

In Europe, on the other hand, the relevant legislation seems to have been drafted with a great deal of optimism for the future. Apparently, the assumption is that the problem of natural monopolies can be cured with a few clear regulations, leaving any other problems for the competition authorities. Not only is it difficult to see how this could be possible, it is also hard to imagine why the European legislator would consider this a good idea. After all, the regulator will normally be in a much better position to understand the market, while the fact that the competition authority will only be able to intervene *ex post* will also cause great uncertainty among market actors, reducing their willingness to invest.

At the very least, this European approach provides a clear answer to our original question. There is certainly a place for competition law in regulated industries, but perhaps legislators on both sides of the Atlantic ought to reconsider what it is.

Notes

1. Case C-280/08 P, *Deutsche Telekom v. Commission* (2010), ECR I-0000, upholding Case T-271/03, *Deutsche Telekom v. Commission* (2008), ECR II-00477.
2. Ibid., par. 77–96.
3. *Pacific Bell Telephone Co. v. LinkLine Communications*, 555 US ___ (2009). 129 S.Ct. 1109, 172 L.Ed.2d 836.
4. Section 601(b)(1) of the Act, codified as 47 USC 152, note.
5. Exceptions include De Streel, A., 'The new concept of 'Significant Market Power' in electronic communications: The hybridisation of the sectoral regulation by competition law', *European Competition Law Review*, vol. 24 (2003) no. 10, pp. 535–42, D. Geradin, 'Limiting the scope of article 82 EC: what can the EU learn from the US Supreme Court's Judgement in *Trinko* in the wake of *Microsoft*, *IMS*, and *Deutsche Telekom?*', *Common Market Law Review*, vol. 41 (2004), pp. 1519–53, G. Psarakis, 'Sector-specific regulation and competition law in the electronic communications sector against the backdrop of the internal market', *European Competition Law Review*, vol. 28 (2007) no. 8, pp. 456–63, P. Alexiadis and M. Cave, 'Regulation and competition law in telecommunications and other network industries', in R. Baldwin, M. Cave and M. Lodge (eds), *The Oxford Handbook of Regulation* (Oxford: Oxford University Press, 2010) pp. 500–22, and D. Hesseling and T. Vermeulen, 'Access to networks through competition law: the case of KPN – Reggefiber', *Network Industries Quarterly*, vol. 13 (2011) no. 1, pp. 14–6.
6. An early non-legal article is S. K. Schmidt, 'Commission activism: subsuming telecommunications and electricity under European competition law', *Journal of European Public Policy*, vol. 5 (1998) no. 1, pp. 169–84.
7. Directives 2002/19, 2002/20, 2002/21 and 2002/22 of 7 March 2002, OJ L 108, pp. 1–77.
8. *Verizon Communications Inc. v. Law Offices of Curtis V. Trinko, LLP*, 540 US 398, 124 S.Ct. 872 (2004).
9. Pub. L. 104–104, 110 Stat. 56, codified in various places in Title 47 of the United States Code.
10. See, for example, *United States v. National Assn. of Securities Dealers, Inc.*, 422 US 694 (1975) and *Gordon v. New York Stock Exchange, Inc.*, 422 US 659 (1975).
11. Section 601(b)(1) of the Act, codified as 47 USC 152, note.
12. *Verizon v. Trinko*, op. cit., at 406, quoting *United States v. National Assn. of Securities Dealers*, op. cit., at 728–30.
13. 15 USC 2.
14. Cf. *Pacific Bell v. LinkLine*, Brief by respondents in opposition to *certiorari*, pp. 30–4. The lower court rulings are *LinkLine Communications v. SBC California*, 2004 WL 5503772 (C.D. Cal.), upheld in *LinkLine Communications v. SBC California*, 503 F.3d 876 (CA 9).
15. Cf. Case C-52/09, *Konkurrensverket v. TeliaSonera Sverige AB* [2011], ECR-I-0000, par. 96–103.

16. *Pacific Bell v. LinkLine*, op. cit, at 11, quoting *Brooke Group Ltd v. Brown & Williamson Tobacco Corp*, 509 US 209, at 222–224 (1993).
17. *Matsushita Elec. Industrial Co. v. Zenith Radio Corp.*, 475 US 574, at 594 (1986).
18. Case C-280/08 P, *Deutsche Telekom v. Commission* (2010), ECR I-0000, upholding Case T-271/03, *Deutsche Telekom v. Commission* (2008), ECR II-00477.
19. Case C-52/09, *Konkurrensverket v. TeliaSonera Sverige AB* (2011), ECR-I-0000.
20. Note, however, Recital 11 of Regulation 2887/2000 on unbundled access to the local loop, (2000) OJ L 336/4: 'Pricing rules for local loops should [...] ensure that there is no distortion of competition, in particular no margin squeeze between prices of wholesale and retail services of the notified operator. In this regard, it is considered important that competition authorities be consulted'.
21. Cf. Joined Cases C-359/95P and C-379/95 P, *Commission and France v. Ladbroke Racing* (1997), ECR I-6265, par. 33–34.
22. Commission Decision 88/518, Napier Brown – British Sugar, (1988) OJ L 284/41.
23. *Industrie des poudres sphériques v Commission*, Case T-97/5, (2000) ECR II-3755, par. 178.
24. Cf. *Deutsche Telekom v. Commission*, op. cit., par. 149–152 of the judgement on appeal, where the appellant tried to argue for the approach taken by the American Supreme Court, and par. 155–185 for the court's argument to the contrary.
25. Most importantly *Akzo Chemie v. Commission*, Case C-62/86, [1991] ECR-I 3359. Note that among the abuses explicitly listed in art. 102 TFEU is 'unfair purchase or selling prices'.
26. *Konkurrensverket v. TeliaSonera Sverige*, op. cit.
27. Cf. *TeliaSonera*, op. cit., par. 75–77.
28. Cf. *Deutsche Telekom v. Commission*, op. cit, par. 250–251 and *TeliaSonera*, op. cit, par. 69–70.
29. *TeliaSonera*, op. cit., par. 104.
30. Aghion, P. and P. Howitt, 'A model of growth through creative destruction', *Econometrica*, vol. 60 (1992) no. 2, pp. 323–51.
31. Cf. Schumpeter, J. A., *Capitalism, Socialism and Democracy* (New York: Harper & Row, 1942), particularly chapter VII.
32. Cf. art. 102(b) TFEU, which explains that abuse of market power may consist of 'limiting [...] technical development to the prejudice of consumers'.
33. Cf. Surowiecki, J., 'Blockbuster, netflix, and the next level', *The New Yorker*, 18 October 2010.
34. Cf. Vezzoso, S., 'The incentives balance test in the EU Microsoft case: a pro-innovation 'economics-based' approach?', *European Competition Law Review*, vol. 27 (2006) no. 7, pp. 382–90.
35. Cf. Stigler, G. J, 'The theory of economic regulation', *Bell Journal of Economics and Management Science*, vol. 2 (1971) no. 1, pp. 3–21 and J.-J Laffont and J. Tirole, 'The politics of government decision-making: a theory of regulatory capture', *Quarterly Journal of Economics*, vol. 106 (1991) no. 4, pp. 1089–127.
36. Cf. *TeliaSonera*, op. cit., par. 22, citing *Roquette Frères*, Case C-94/00, (2002) ECR I-9011, par. 42.

37. *TeliaSonera*, op. cit., par. 108–111. Cf. J. Simpson and A. L. Wickelgren, 'Bundled discounts, leverage theory, and downstream competition', *American Law and Economics Review*, vol. 9 (2007) no. 2, pp. 370–83.
38. *TeliaSonera*, op. cit., par. 109.
39. On 25 March 2011 it was announced that the Consumer Authority and the Mail and Telecom Regulator OPTA were similarly going to be brought under the umbrella of the NMa. While the details of this merger still have to be worked out, it is not expected that the outcome will be anything other than two more 'chambers' within the NMa.
40. Cf. Hendrikse, G., *Economics and Management of Organizations* (New York: McGraw-Hill, 2003), Ch. 6.
41. Cf. NMagazine December 2008, pp. 8–9, 'Twee helikopters boven dezelfde markt'.
42. At this point we will not attempt to formalise any difference in the ability of each agency to acquire knowledge.
43. The assumption that the discovery occurs with certainty is without loss of generality, unless we wish to assume that the company is risk averse.
44. Cf. Gintis, H., *Game Theory for Behavioral Scientists* (Princeton: Princeton University Press, 2007), Ch. 6.
45. This expression is equivalent to the one given above, and still assumes that $\pi(B)=0$.
46. Op. cit., cf. section 3, above.
47. Directives 2002/19, 2002/20, 2002/21 and 2002/22 of 7 March, (2002) OJ L 108, pp. 1–77.
48. Cf. Alexiadis and Cave (2010), op. cit.
49. COM (2000) 384, p. 2.
50. Directive 2002/19 on access to, and interconnection of, electronic communications networks and associated facilities (Access Directive), (2002) OJ L 108/7.
51. COM (2000) 384, p. 3.
52. Commission Decision C(2003) 1536, *Deutsche Telekom* (2003), OJ L 263/9.
53. Cf. Commission Decision C(2003) 1536, op. cit., par. 34–45.
54. The most famous such use of a structural measure is of course the case of *United States v. AT&T*, 552 F.Supp 131 (D.C. DC, 1982), where competition law was used to break up AT&T into seven 'Baby Bells'.
55. Cf. Pollitt, C. and G. Bouckaert, *Public Management Reform: A Comparative Analysis* (Oxford: Oxford University Press, 2004), Ch. 7.
56. Cf. *Verizon v. Trinko*, 504 US 398, at 407 and Geradin (2004).

7
Is Europe Turning into a 'Technological Fortress'? Innovation and Technology for the Management of EU's External Borders: Reflections on FRONTEX and EUROSUR

Luisa Marin

7.1 Once upon a time …

Once upon a time, in an era where the states defined themselves as 'sovereign', there was a border point. The border point was a place where a guard or a police-like officer used to ask to check a document called passport, which entitled a person to cross the border. States' ambitions to enhance controls over the flux of non-citizens entering their territories led them to raise the requirements and ask for an extra document, the visa, a form of permission required before arriving at a state's port or entry.[1]

Nowadays the social physiognomy of the border has radically changed for a wide range of reasons. First, we name the context where the border performs its function.

Looking at the European region, the internal market project and its freedom of movement *rationale* have deprived member states' (hereinafter: MS) borders of most of their meanings. Lately, the Schengen process has removed controls at internal frontiers and required the strengthening of external borders for the benefit of European citizens.[2] The ambitious project of the Area of Freedom, Security and Justice (hereinafter: AFSJ), 'without internal frontiers, in which the free movement of persons is ensured in conjunction with appropriate measures with respect to external border control, asylum, immigration, and the prevention and combating of crime'[3] has consolidated these milestones into something aimed at being a more coherent and comprehensive project.[4]

At another level, globalisation is challenging the purpose of controlling the flux of individuals moving across the globe, a single spatial and geo-political entity, a 'global village';[5] at the same time, states are faced with the increased phenomena of mobility by people coming from disadvantaged areas of the world. Persistent poverty, recently made more acute by the economic crisis, is exasperating this situation, as recent North African political turmoil demonstrates.[6] In this context, states are trying to strengthen their controls over migration flows, in an attempt to manage pressures and challenges for domestic economic and welfare systems.

A second factor to be highlighted while attempting to sketch how border controls have changed is technology.[7] Nowadays the discourse is about e-borders, digital borders[8] or technological borders. The myriad of technological applications available, from iris and bone scans, to satellites, 'drones' – Unmanned Aerial Vehicles (UAV) – and databases, has changed the nature of controls performed at borders.[9] The term 'smart borders' suggests that controls at borders are becoming more and more ingenious, thanks to technology.[10]

European integration (with its Schengen spin-off) and globalisation, on the one hand, and the spread of technology in the management of border control, on the other hand, have determined the functional metamorphosis of the EU's external borders: these now resemble the gates of a 'cyber-fortress',[11] erected by Europe together with its MSs in an effort to control the endless migration flows approaching the shores of the EU.

External borders, even if naturally porous like sea borders, are framed in the political discourse as gateways to risks and threats that the EU and its MSs need to manage and control, such as cross-border crime, terrorism and 'illegal' migration. This is happening in parallel with the securitisation of migration control, which also affects the nature of border control, and which can be considered now as a fully fledged policing activity.[12] Recently commercial actors, such as aerial companies, have also been involved in the practices of controlling borders, as states' agencies delegate document control to them. The emergence of these phenomena is conceptualised as 'policing at distance', 'remote controls' and 'externationalisation of migration':[13] all these theories suggest that border control is getting an outreach dimension, purely instrumental to migration management,[14] or, in the words of Elspeth Guild, that borders are 'moving'.[15]

The aim of this chapter is to examine critically the latest developments on technologies for the purpose of border control at EU and MS levels. FRONTEX, the European Borders Agency, one of the last cases of agencification in the AFSJ, represents an institutional reform with the purpose of coordinating and thus 'putting together' brute force technologies (e.g. helicopters and vessels) usually 'belonging' to domestic military agencies. The second case examined under the heading of technology is EUROSUR, the European Border Surveillance System, which is developing within an ever-growing plethora of technological systems, infrastructures and databases serving the purpose of surveillance and management of external borders.[16] The third case aims to present current researches and 'works in progress' in the field of technologies for the surveillance of external borders, such as 'drones' or UAV, which might be informative as to the direction in which cooperation in border control is developing.

These examples provide prisms though which we can look at how innovation and technology are translated in the context of policing external borders, and assess their implications in a political and legal perspective.

FRONTEX and EUROSUR must be placed into the EU's policy and legal context, that is, against the framework of the Stockholm programme and of the Lisbon Treaty. The Stockholm programme[17] is the last multiannual programme of the EU for the AFSJ. Both texts provide new impetus to the AFSJ, defining new policy plans within a new legal architecture. In particular, the multiannual programme aims to offer an open and secure Europe serving and protecting the citizens, where openness and security represent two antithetic, even conflicting, paradigms inspiring European policies.[18]

In the Stockholm programme border and migration control are given high importance, both in their internal and external dimensions, inasmuch as it is possible to distinguish between them.[19] Under the heading of 'access to Europe in a globalized world' the EU aims to strengthen the integrated borders management and visa policies. Illegal migration and cross-border crime are framed as phenomena which need to be contrasted, and FRONTEX is designated a central role in this respect, together with EUROSUR.[20]

Furthermore, the programme confirms and builds upon the Global Approach to Migration, which was initiated in 2005 and centres on three axes: promoting legal migration, the relation between migration and development, and the fight against illegal migration. While recalling the advantages and chances represented by increased mobility and migration for the EU's MSs, and the need for MSs to adopt pro-active policies establishing links with the national labour market requirements, the programme stresses that the fight against illegal migration needs to be developed through the integrated border management system, as well as cooperation with the country of origin and transit, completed also by return policies.[21] The programme shows how the EU's migration policy has already acquired an outreach dimension and is expanding increasingly toward cooperation with third countries. If the programme is crowned by ambitious words and references to European values, it remains to be seen whether the AFSJ's practice matches its declarations of principles and promises of rights.[22] For example, the programme reveals that border controls are purely instrumental to tackling migration; states' attempt to control migration implies also a criminalisation of the phenomenon, without considering its causes, and, even more importantly, it means that EU MSs are not refraining from cooperating with, for example, North African illiberal regimes, which could have the adverse effect of increasing migration toward the EU.

Having presented the context and function of border control and summarised the recent legal and political framework (1), this chapter will proceed as follows: the next section will be devoted to the main institutional innovation in the field, that is, the creation of the European agency FRONTEX, explaining its purpose, legal framework and some of its operations and issues associated with them (2). The next section will present the technological system for the management and surveillance of the external borders, EUROSUR (3). A fourth section will inform about current research and cooperation projects in the field of border surveillance (4), showing a trend towards militarisation of border surveillance. The paper will conclude (5) with some considerations on the political implications of these choices and developments, putting forward the argument that turning Europe into a technological fortress will not benefit Europe itself, nor its inhabitants.

7.2   FRONTEX: The agency and (some of) its operations at sea

The European Borders Agency FRONTEX (from *Frontières extérieures*) was set up in 2004[23] and reformed in 2007 (hereinafter: FRONTEX Regulation and RABIT Regulation, respectively).[24]

The agency was created as a Community Agency,[25] in line with the consolidated case law on implied powers:[26] its legal basis is to be found in the old (pre-Lisbon) Treaty of the European Community (TEC), namely Article 62(2)(a) and Article 66 TEC, granting the EC powers to adopt measures on the crossing of the external borders by establishing standards and procedures to be followed by MSs in carrying out checks on persons at such borders and measures to ensure cooperation between the relevant departments of the MSs' administrations and between Commission and MSs.[27]

Article 1 of the FRONTEX Regulation states that FRONTEX was established with the mission of 'improving the integrated management of the external borders of the Member States'; the concept is defined in a Council document of late 2006,[28] endorsed by the European Council on 4–5 December 2006,[29] and comprises:

1. Border control (checks and surveillance);
2. Detection and investigation of cross-border crime;
3. The four-tier access control model, comprising measures in third countries (hereinafter TC), cooperation with neighbouring countries, border control, and control measures within the area of movements, including return;
4. Inter-agency cooperation;
5. Coordination and coherence on actions at EU level.

It appears that border checks and surveillance are only the first point of a much longer list, which also comprises investigation of cross-border crime and cooperation with TCs.[30]

The legal text regulating the internal and external border management, the Schengen Border Code,[31] was adopted later in 2006, more than one year after FRONTEX began operating.[32] In this respect, FRONTEX offers an example of the dynamic and unstable relation between law and politics: in some circumstances political pressure pushes forward projects before the overall and necessary legal infrastructures are in place.

Among the main reasons for setting up a new agency we have 'provide the Commission and the MS with the necessary technical support and expertise in the management of external borders'.[33]

According to the FRONTEX Regulation, the agency's main tasks are to:

1. coordinate operational cooperation between MSs in the management of the EU's external borders;[34]
2. assist MSs in circumstances requiring increased technical and operational assistance at external borders;[35]
3. provide MSs with the necessary support in organising joint return operations.[36]

Besides this core operational dimension, the other main tasks include:

1. assisting MSs to train national border guards, including through the establishment of common training standards;
2. carrying out risk analyses;
3. following up on the development of relevant research for the control and surveillance of external borders.[37]

The agency is required to assess, approve and coordinate joint operations and pilot projects proposed by MSs; FRONTEX itself may launch such operations, and is also empowered to put its technical equipment at the disposal of the MS joining such operations, as well as offer financial backing.[38]

Other provisions show that FRONTEX plays an important role in 'putting together' technical resources among MSs, which witnesses the solidarity between MSs. The agency shall set up and keep a centralised record of MSs' technical means and equipment for the control and surveillance of external borders; MSs should contribute on a voluntary basis and according to the needs of the requesting MS.[39]

The reform of 2007 was aimed at providing a rapid crisis-response capability available to all MSs, through so-called Rapid Border Intervention Teams (RABITs). This is the additional mission of FRONTEX, the tool to be used when risk analysis and intelligence activities (also by FRONTEX) fail to predict risks or events that MSs need to react to.[40] The RABITs should provide support for a limited time in exceptional and urgent situations, such as 'mass influx of third-country nationals attempting to enter a member state's territory illegally'.[41]

Another provision of the RABIT Regulation worth considering is that concerning the tasks and powers of members of teams: Article 6 specifies that team members shall have the powers to achieve border checks and border surveillance, and, more generally, the objectives of the Schengen Border Code, in accordance with the operational plan specified before the start of the RABIT intervention. This means that team members shall be authorised to carry service weapons, ammunition and equipment, and also to use them in accordance with the law of the MS hosting the RABIT intervention. The host MS shall also specify before the start of the mission which weapons shall be authorised and the conditions of their use. In any case, team members shall be authorised to use weapons for legitimate self-defence and legitimate defence of the members of the teams or other persons.[42]

The FRONTEX Regulation provides for the organisation and realisation of the main tasks presented above. Other provisions of the FRONTEX Regulation are worth considering because of the scope of action they allow to the Agency.

For example, Article 13 enables FRONTEX to cooperate with EUROPOL and other international organisations competent in areas covered by the FRONTEX Regulation, through 'working arrangements', whereas Article 14 establishes some 'external capacities' with TCs and their authorities: indeed the agency 'shall facilitate the operational cooperation between Member States and third countries, in the framework of the European Union external relations policy', through facilitation agreements. Besides this, the agency can conclude 'working arrangements' with competent TC's authorities. Summing up, FRONTEX can conclude two different types of external agreements: (a) 'facilitation agreements' with TCs, and (b) 'working arrangements' with TC's relevant authorities, which also apply to inter-agency cooperation.[43]

Analysing FRONTEX functioning, a first assessment to be made is that the core focus of the agency is on operational aspects: coordinating MSs' operational cooperation and providing assistance to their authorities.[44] The reform of 2007 confirmed this trend, strengthening the tasks and powers of officers participating in FRONTEX operations:[45] there seems to be a stable evolution stressing the operational dimension.[46] This is also the way the agency profiles itself. Its reports are written in a very technocratic jargon, stressing cooperation aspects, coordination functions and management logics applied to external borders. All this is meant to evocate knowledge and expertise-based legitimacy, and output legitimacy. This in turn is intended to strengthen the credibility of its operations, which constitute the main expenditure of the FRONTEX annual budgets.[47]

An overall analysis of FRONTEX Joint Operations (hereinafter: JO) at sea borders would be beyond the remit of this chapter: therefore we will present here some of those JOs carried out at the Western and central Southern maritime border in order to gain a better understanding of the way that technology and technical cooperation affect border controls and surveillance.

In JO HERA (in particular HERA II and III), MSs were cooperating with Spain under the coordination of FRONTEX in joint sea surveillance operations, that is, joint aerial and naval patrols with the purpose of intercepting and diverting boats, in cooperation with the authorities of Senegal and Mauritania, the states of departure for these 'boat peoples'. In a FRONTEX News Release on JO HERA, one can read that the agency detected vessels setting off toward Canary Islands, and diverted them back to Senegal and Mauritania[48] on the basis of bilateral agreements between Spain and Senegal and Spain and Mauritania.

JO NAUTILUS, based on the central Mediterranean Sea, began in 2006.[49] JO NAUTILUS IV (2009) is especially problematic, for several reasons. First, there is little official information by FRONTEX as to what has happened at sea during operations NAUTILUS I, II and III and no information at all on NAUTILUS IV.[50] Secondly, the information by other sources, such as NGOs and academic reports about NAUTILUS II offer conflicting statements by Maltese officials as to whether push-backs of 700 migrants to Libya occurred.[51] Besides this, one can learn that the Schengen Border Code was not applied because Malta was not yet a member of the Schengen Agreement. The legal basis of those operations was unclear, and how to intercept migrants and where and how to take them were discussed on an ad hoc basis by military and security officials, thus 'reinforcing [...] [their] discretion'.[52]

The NGO Human Rights Watch[53] denounced NAUTILUS IV, stating that it 'resulted in the interdiction and push back of migrants in the central Mediterranean Sea to Libya', with the cooperation of a German *Puma* helicopter, under the coordination of the Italian coastguard. The boat was carrying 75 migrants and has been 'handed over' to a Libyan patrol boat, which 'took them' to Tripoli, where they were assigned to a military unit. The Human Rights Watch report also quotes a declaration by FRONTEX Vice-Director Gil Arias-Fernandez, who commented favourably on this operation: 'Based on our statistics, we are able to say that the agreements [between Libya and Italy] have had a positive impact. On the humanitarian level, fewer lives have been put at risk, due to fewer departures. But our agency does not have the ability to confirm if the right to request asylum as well as other human rights are being respected in Libya'. The FRONTEX director denied the involvement of the agency in push-backs, and clarified that Italian operations took place outside FRONTEX operational area.[54] Lacking any official information by FRONTEX as to the operational area, it is, as first, hard to get a clear picture of what has happened, and, secondly, difficult to believe that there was no contiguity at all between a FRONTEX JO and a single MS's initiative.

As to FRONTEX, one should observe that the coordination façade of this agency and the technocratic jargon of its reports are an attempt to hide a reality of technical militarisation of border control and surveillance: also thanks to FRONTEX, these activities are now performed with a rich deployment of several types of military equipment, ranging from aerial and naval crafts to weapons. The domestic agencies involved, such as the Italian *Guardia di Finanza*, have a quasi-military status in their domestic administrative systems. One might legitimately wonder if all this deployment of military force against undocumented migrants, implying a significant expenditure of public money, is desirable. In spite of the increasing political pressure to tackle migration flows coming from African shores, academic analysis show that irregular migration at maritime borders does not represent a significant *ratio* of the whole phenomenon, and that the successful effects of operations at sea is not clear.[55] This requires attention because it undermines the political desirability of these type of initiatives, and raises questions as to the proportionality between resources involved and final results achieved.

The little official information offered by FRONTEX does not meet the standards of accountability and transparency with which the agency should comply. Secondly, the migration dimension of these operations, and thus, the impact on human lives, is completely neglected in the agency's assessment. These operations might expose migrants to longer and more dangerous sea travels, if they disrupt shorter routes.[56] Reading FRONTEX reports it appears that the agency's only objective is erecting borders against a 'criminal phenomenon', illegal migration, and it does not matter that this concerns desperate people, often women and children, seeking a new place to live at risk of their own lives.

Besides this, there are also other legal questions that undermine the legality of FRONTEX operations: for example JOs HERA II and III were carried out on the basis of bilateral agreements between Spain and TCs that have not been made public: the EU (or FRONTEX) and participating MSs were not part of those agreements.[57] Italian push-backs that occurred alongside NAUTILUS IV were made possible on the basis of bilateral agreements between Italy and Libya.[58] Secondly, the operations presented in this section should be framed as diversion or interception operations against migrants, some of whom are potential asylum seekers: after concerned reactions by many NGOs and international agencies, such as the UN High Commissioner for Refugees,[59] the academic literature analysed the legality of those operations with reference to the law of the seas, international human rights and refugee law, with a special emphasis on the principle of *non refoulement*, as well as European asylum law, putting forward many criticisms.[60]

The relevance of these questions became more serious after the entry into force of the Lisbon Treaty: the legally binding nature of the Charter of Fundamental Rights of the EU, with its many references to international law instruments, such as the Geneva Convention on the Status of Refugees, is going to make the problem more acute. A fragmentary interpretation of MSs' and FRONTEX' duties while monitoring the EU's external borders is no longer sustainable. Thanks to the broader scrutiny of the EU Court of Justice,[61] the legal accountability mechanisms will make it more likely that current operations will be assessed.

7.3   EUROSUR: The EU's technological system for the control and surveillance of external borders

Since the Hague Programme the EU political agenda has been developing in the direction of exploiting all the possibilities offered by technologies within the realm of the policies falling within the AFSJ: WE refer, for example, to the principle of availability, meaning that 'a law enforcement officer in one Member State who needs information [in the pre-trial phase] in order to perform his duties can obtain this from another Member State and that the law enforcement agency in the other Member State which holds this information will make it available for the stated purpose'.[62]

Technology is already a tangible reality thanks to a number of databases. Among the systems in place for the AFSJ's policies are the SIS (the Schengen Information System), which is designed to map persons who should be refused entry or be searched by law enforcement authorities, and the EURODAC systems, which lists asylum seekers. More systems are currently (and still) being developed: the SIS II, the VIS, for travellers requiring visas, and the E/ES, the Entry/Exit System, for so-called overstayers, the most significant group among irregular migrants. These have been explained as 'technological' or 'digital fixes' for the EU.[63]

In the field of management and surveillance of the EU's external borders EUROSUR is currently being created: the target for 2011 is the EUROSUR Pilot Project, which is 'developing and demonstrating the exchange of relevant information between the Members States as well as between the Member States and Frontex'. It will first be operational between six countries at the Southern and Eastern Borders. Its aim is to 'support MSs by developing systems with modern technologies, by promoting interoperability and uniform border surveillance standards and by extending cooperation and improving data sharing between Member States and Frontex'. [64] Other relevant contributions of FRONTEX to EUROSUR in 2011 are the development of a Common Pre-frontier Intelligence Picture, to be realised by the Risk Analysis Unit; and the exploration of satellite-based imagery for border security.[65]

The EUROSUR project originates from two feasibility studies the Council commissioned FRONTEX to undertake: MEDSEA and BORTEC.

The studies are not public although there is some information on them in other official documents. The MEDSEA study was designed to explore reinforcing the monitoring and surveillance of the Mediterranean, through a Coastal Patrols Network, involving Southern European countries and North African states. The BORTEC report has mapped the situation in place with regard to border surveillance, in order to explore the feasibility of a comprehensive borders surveillance system: the results showed that about 50 authorities from 30 institutions are involved in border surveillance, sometimes with overlapping competences and systems.[66]

As a follow-up to those studies, the Commission Communication of 2008[67] presented the challenges and objectives for the future development of border surveillance. In that text the EUROSUR project was meant to focus initially on the EU's Southern and Eastern borders. Its declared objectives were to reduce the number of illegal migrants entering the EU; to increase the security of the EU as a whole by contributing to the prevention of cross-border crime; thirdly, to enhance the EU's search and rescue capacity, an objective which in later texts has been framed as a humanitarian reason for reducing the death toll of migrants losing their lives while attempting to cross the Mediterranean.[68]

The EU's goal with EUROSUR is to achieve 'border surveillance', which means to set up a system for the surveillance of borders between crossing points. This should be completed by 'border checks', that is, border control through checks carried out at border crossing points, which is fulfilled by technological tools currently being developed (the abovementioned SIS II, VIS, E/ES, and the RTP, the Registered Traveller Programme). Border surveillance through EUROSUR is indicated to represent an important step in the process of gradual establishment of a common European integrated border management system.

In practical terms, EUROSUR is a common technical framework to support MSs authorities and enhance their capacity to coordinate at a European level and to cooperate with TCs. Thus, the main goal of EUROSUR is to provide a technical framework for the use of existing systems and common tools, like satellites, and for the exchange of information and intelligence. The EUROSUR system should be set up without affecting MSs' respective areas of jurisdiction (EUROSUR's principle of 'neutrality' with respect to internal MSs' division of competences), nor by replacing any existing systems.

Its development should take place through three phases: the first phase is about interlinking and streamlining existing national surveillance systems at MS level; the second, about the development and implementation of common tools at the EU level; the third phase should lead to a common information-sharing environment for the EU maritime domain.

In the first phase MSs will have to set up National Coordination Centres (NCCs) and National Surveillance Systems (NSSs). This shows EUROSUR envisages the creation of new actors tasked with intelligence functions at the domestic level. At the same time, other steps of the first phase require some cooperation with third countries. This last aspect is having deep political implications, and therefore, should be dealt with at the appropriate level, with guarantees of transparency.

More recently, in the 2010 Communication on 'The EU Internal Security Strategy in Action: Five steps towards a more secure Europe'[69] the Commission addressed the EU strategy on security through five policy objectives, confirming that one of the key objectives for addressing security concerns was to strengthen border management, a target to be reached mainly thanks to technology.

If the legitimacy of EUROSUR is defended also for its humanitarian dimension, one might wonder whether all these developments are not driven simply by a logic of surveillance and control of migration phenomena, to be carried out by exploiting all the possibilities offered by the high-tech industry.

The EU is investing significant amounts of money in technologies for policing borders:[70] EUROSUR is part of this process. While the rationalising effort of this project is laudable, its long-term implications are more obscure, especially with regard to control over the amount of data to be generated by all the databases indicated and the flow of information to be exchanged though the system, the magic word being interoperability. Secondly, any expansion of controls over migration must be confronted with the so-called waterbed effect: though it is often reported that recent increased practices of controls and surveillance at sea have contributed to diminished numbers of illegal migrants, these arguments do not consider that increased controls impact on migration routes more than on migration itself. Moreover, if the number of people drowning at sea falls,[71] the numbers of missing people thanks to sea surveillance and interception practices is rising.[72]

7.4   Other investments in brute-force technology

The panorama of European developments and investments in brute-force technology does not stop here. Among other plans for research on hi-tech border surveillance are satellite surveillance systems, UAVs or 'drones', and maritime vehicles, and other systems to be studied in the context of border management. The flow of technological tools and systems to be exploited to defend Europe from the 'threat' of undocumented migrants seems never-ending.

Also in this context, FRONTEX is cooperating with defence and surveillance industry, in the context of the PF7 programme, working to adapt military surveillance techniques for Europe's borders. The 'giants' of war and military industry, such as Sagem, Finmeccanica, Israel Aircraft Industries and others are working together within EU-sponsored research and industry projects to develop what we would label 'total surveillance tools', that is, tools and technical infrastructure to achieve the '24/7 blue and green border situation awareness'.[73]

The EU is thus funding research aiming to develop autonomous and adaptive systems for protecting borders, taking measures to stop illegal action at those borders with the supervision of border guards. FRONTEX is investing in research on fixed surveillance and border-drone technology, alongside similar projects studied by the European Defence Agency (EDA) and some MSs. Besides this, it is also coordinating a working group of the European Security Research and Innovation Forum (ESRIF), dealing with integrated border management and maritime surveillance.[74] The US has been also investing in such technologies for controlling its Mexican borders through 'predator drones', which have been heavily criticised by UN Special Rapporteur on Extra-judicial Killings Philip Alston, who accused the US of 'giving the CIA a license to kill and encouraging a PlayStation mentality that devalues human life'.[75] Even accepting the Commission's argument that the EU is developing such technological tools with the aim of reducing the death toll of migrants, a question remains: will such a plethora of technological tools tell us more than mere facts, that is, the presence of humans in the midst of the sea? Will this technology help FRONTEX and the MSs to identify those people trying to reach the shores of Europe, in fulfilment of international and European human rights obligations? Or are these systems meant to reinforce 'cyber-fortress Europe'?

Another aspect emerging in reports of FRONTEX's high-ranking officials[76] concerns involvement and cooperation with armed forces in the control of migration flows. The ongoing cooperation with NATO and EDA as well as with national armed forces in the area of border security issues[77] reveals a process of militarisation of border controls. Even if this is confined to the area of research and technology, training and exchange of information and risk analysis, there are reasons to believe that technology is becoming the 'Trojan horse' for pushing forward more militarisation of border surveillance within a logic of securitisation of borders management and migration policy.[78] The enemy that our democratic and liberal societies are facing is the un-wealthy and un-documented migrant, trying to reach his or her ought-to-be 'promised land' in precarious boats: we are tackling this threat by deploying satellites, airplanes, vessels, sensors and more, as if we are dealing with enemy states – or 'rogue states' such as Libya –, jeopardising the very existence of our democracies and liberal values. Most worrying of all, the EU and its MSs are actively involving non-liberal states in this fight against migrants, by making cooperation and aid subject to 'migration conditionality'.

The rationale underlying the creation of EUROSUR, as emerging from the Commission Communication of 2008, reveals a similar logic, while framing a close link between the need to fight illegal migration and prevent cross-border crime (terrorism, trafficking in human beings, drug smuggling and illicit arms trafficking).[79] These developments need to be monitored simply because they confirm the criminalisation of undocumented migration. They thus suggest that undocumented migrants should simply be stopped before the borders, without consideration for any other elements which might come into play, like people fleeing persecution and the like.

7.5 Concluding remarks

After an introduction putting the EU's external borders control in the context of the fight against undocumented migration, the chapter has presented the legal framework and some of the activities of the EU's border agency FRONTEX, focusing in particular on maritime border control in the Mediterranean Sea. The information available on some JOs (described above) indicates serious policy and legal issues on the respect of international and European asylum provisions, besides problems of transparency and accountability on such activities led by FRONTEX. In particular, the interception of boats of undocumented migrants at sea, carried out within or alongside FRONTEX-coordinated operations, constitute a deplorable practice. The institutional innovation represented by FRONTEX will be completed by the EUROSUR system, currently under discussion. The latter instrument has the purpose of realising a 24/7 control of the EU's land/sea borders, in order to achieve 'full situational awareness'. Borders control is – or could be in the near future – performed with the deployment of all technological means available, from biometrics to war technologies (drones). Considering the type of activities (interception of boats at sea) and the instruments employed or being investigated/developed, that is, in the direction of military technologies, the first consideration to be made is about the ongoing militarisation and securitisation of external borders control and surveillance. The question here is whether pursuing such a policy is socially desirable, having regard to the high costs involved and the unclear benefits.

A second reflection considers that pursuing the path of technology without clearly defined parameters of what is socially desirable and what is not, undermines the whole high-tech project on issues of legitimacy and proportionality. Technology runs the risk of being exploited to achieve 'total surveillance' situations, which would threaten our liberal and democratic societies much more than some thousands third country nationals in search of peace and prosperity. In this scenario technology, a mean, becomes an end.

Another concern is devoted to persons, as stakeholders affected by border control and surveillance. The interests and aspirations of persons seeking to reach Europe force us to reconsider the way we frame the relation between person and territory first, but also between we Europeans and the others, that is, persons coming from the rest of the world. Making access to Europe more difficult will not stop people in distress from searching for more humane conditions in which to live; quite the opposite, it might imply that more persons will fall in smugglers' hands and more generally fall victim to organised crime. Therefore, one should be aware that the way in which access to a country is framed has implications for the freedom of other persons, for the notion of a person as a holder of rights, and eventually for the very basic idea of what constitutes a human being.[80]

Turning Europe into a technological fortress and disregarding human life in its bare life – *nuda vita* – dimension will not make Europe and its privileged inhabitants safer. The developments presented in this chapter should indicate the need for a reflection on the social desirability of the massive deployment of technologies in the domain of border control and surveillance. We should not forget that borders should define the place where they *all* lived happily ever after.

Notes

1. Ryan, B., 'Extraterritorial immigration control: what role for legal guarantees?' in B. Ryan and V. Mitsilegas (eds), *Extraterritorial Immigration Control. Legal Challenges* (Leiden and Boston: Martinus Nijhoff 2010), p. 4.

2. See Guild, E., *Moving the Borders of Europe*, inaugural lecture, available at: www.jur.ru.nl/cmr/docs/oratie.eg.pdf; see also K. Groenendijk, E. Guild and P. Minderhoud (eds), *In Search of Europe's Borders* (The Hague: Kluwer, 2003).

3. Article 3(2) TEU.

4. The status quo however is often perceived as problematic, unbalancing (internal) security and freedoms, focusing on cooperation among state authorities and undermining individuals' rights.

5. The reference is to the fortunate theorisation of the communication theorist M. McLuhan, *The Gutenberg Galaxy: The Making of Typographic Man* (Toronto: University of Toronto Press, 1962), and *Understanding Media* (New York: McGraw-Hill, 1964)

6. Sapelli, G., 'Disoccupazione, rivolte e immigrazione', *Corriere della Sera*, 19 February 2011.

7. Dijstelbloem, H. and A. Meijer, *Migration and the New Technological Borders of Europe* (Basingstoke: Palgrave Macmillan, 2011).

8. Brouwer, E., *Digital Borders and Real Rights: Effective Remedies for Third-Country Nationals in the Schengen Information System* (Leiden and Boston: Martinus Nijhoff, 2008).

9. In April 2011 the Swiss authorities were patrolling parts of their Southern border with drones, for the purpose of controlling and preventing possible migrants trying to reach the Confederation from Italy. Available at http://www.ilgiorno.it/varese/cronaca/2011/04/07/486460-caccia_migranti.shtml.

10. Dijstelbloem, H., A. Meijer and M. Besters, 'The migration machine, in H. Dijstelbloem and A. Meijer, *Migration and the New Technological Borders*, cit., p. 11.

11. Guild, E., S. Carrera and F. Geyer, *The Commission's New Border Package. Does It Take Us One Step Closer to a 'Cyber-Fortress Europe'?*, CEPS Policy Brief, No. 154/2008.

12. The online version of the Oxford Advanced Learner's Dictionary, for the entry 'policing', reports the following:

    1: the activity of keeping order in a place with police, *community policing*
    2: the activity of controlling an industry, an activity, etc. to make sure that people obey the rules, *the policing of legislation'*.

13. These practices emerged in the context of policing external borders in connection with the fight against undocumented migration. See E. Guild and D. Bigo, 'The transformation of European border controls, in B. Ryan and V. Mitsilegas (eds), *Extraterritorial Immigration Control: Legal Challenges* (Leiden and Boston: Martinus Nijhoff, 2010), pp. 258–9.

14. Guiraudon, V., 'Before the EU Border: Remote Control of the 'Huddled Masses'', in K. Groenendijk, E. Guild and P. Minderhoud (eds), *In Search of Europe's Borders* (The Hague: Kluwer, 2003).

15. Guild, E., *Moving the Borders of Europe*, op. cit.

16. On surveillance see M. Foucault, *Discipline and Punish: The Birth of the Prison* (New York: Vintage Books, 1995). Playing on the concept of Bentham's panopticon, see B. Hayes, *NeoConOpticon: The EU Security-Industrial Complex*. Available at http://www.statewatch.org/analyses/neoconopticon-report.pdf, date of retrieval: 4 April 2011. Transnational Institute and Statewatch, 2009.

17. OJ 2010/C 115/01.
18. For a comment on the draft programme, see E. Guild, S. Carrera and A. Faure Atger, 'Challenges and Prospects for the EU's Area of Freedom, Security and Justice, CEPS Working Document No. 313/2009. See also R. A. Wessel, L. Marin and C. Matera, 'The external dimension of the EU's area of freedom, security and justice', in C. Eckes and Th. Konstadinides (eds), *Crime within the Area of Freedom, Security and Justice: A European Public Order* (Cambridge: Cambridge University Press, 2011), pp. 272–300.
19. It is indeed since 2005 that the European Council has put the milestones for a so-called Global Approach to Migration, witnessing a policy choice of dealing with migration also in the external relations of the EU. See Presidency Conclusions on the Global Approach to Migration: Priority actions focusing on Africa and the Mediterranean, European Council, Brussels, 15–16 December 2005.
20. Stockholm Programme, section 5: Access to Europe in a globalised world.
21. Stockholm Programme, section 6: A Europe of responsibility, solidarity and partnership in migration and asylum matters.
22. The Stockholm Programme features numerous references to fundamental rights and European values through the whole text.
23. Council Regulation (EC) No. 2007/2004 of 26 October 2004 establishing a European Agency for the Management of Operational Cooperation at the External Borders of the Member States of the European Union, OJ L 349/1; hereinafter: FRONTEX Regulation.
24. Regulation (EC) No. 863/2007 of the European Parliament and of the Council of 11 July 2007 establishing a mechanism for the creation of Rapid Border Intervention Teams and amending Council Regulation (EC) No. 2007/2004 as regards that mechanism and regulating the tasks and powers of guest officers; OJ L199/30; hereinafter: RABIT Regulation.
25. Curtin, D., *Executive Power of the European Union Law, Practices, and the Living Constitution* (Oxford: Oxford University Press, 2009), pp. 146, 147, describes agencies as the 'satellite executive power'.
26. The reference is of course to the seminal ECJ's case *AETR*: case 22/70, *Commission v. Council* (AETR) (1971) ECR 263. *See also* J.J. Rijpma and M. Cremona, The extra-territorialisation of EU migration policies and the rule of law, EUI Working Papers LAW 2007/01, quoting readmission agreements adopted on the legal basis of Article 63(3)(b) TEC as an example of act adopted by the EC without an express treaty reference to such instruments, pp. 10–11.
27. In (old) TEC, one reads:
Article 62
'The Council, acting in accordance with the procedure referred to in Article 67, shall, within a period of five years after the entry into force of the Treaty of Amsterdam, adopt:
1. measures with a view to ensuring, in compliance with Article 14, the absence of any controls on persons, be they citizens of the Union or nationals of third countries, when crossing internal borders;
2. measures on the crossing of the external borders of the Member States which shall establish:
(a) standards and procedures to be followed by Member States in carrying out checks on persons at such borders; (...)'.

Article 66 TEC:
'The Council, acting in accordance with the procedure referred to in Article 67, shall take measures to ensure cooperation between the relevant departments of the administrations of the Member States in the areas covered by this title, as well as between those departments and the Commission'.

28. Council document No. 14202/06, draft Council conclusions on integrated border management. The concept has been previously referred at in Commission's communication – toward integrated management of the external borders of the Member States of the European Union, COM (2002)233 final, cit.

29. Council Conclusions on Justice and Home Affairs Council, Brussels, 4–5 December 2006.

30. See Baldaccini, A., 'Extraterritorial border controls in the EU: the tole of FRONTEX in operations at sea', in B. Ryan and V. Mitsilegas (eds), *Extraterritorial Immigration Control. Legal Challenges* (Leiden and Boston: Martinus Nijhoff, 2010), p. 233.

31. Regulation (EC) No. 562/2006 of the European Parliament and of the Council, OJ L 105,01.

32. Guild, E. and D. Bigo, The Transformation of European Border Controls, cit., 268.

33. Article 1(3) of the FRONTEX Regulation.

34. Article 2(1) letter (a).

35. Ibid., letter (e).

36. Ibid., letter (f).

37. Article 2(1) letters (b), (c), (d) respectively.

38. Article 3(1) and (4).

39. Article 7.

40. Regulation No. 863/2007/EC.

41. Ibid., recital 7 of the preamble.

42. Article 6 (1), 6(5), 6(6), 6(7).

43. Cf Wessel, R. A., L. Marin and C. Matera, 'The external dimension of the EU's area of freedom, security and justice', in C. Eckes and Th. Konstadinides (eds), *Crime within the Area of Freedom, Security and Justice: A European Public Order* (Cambridge: Cambridge University Press, 2010), p. 294.

44. This is what one can read in FRONTEX Press release: 'FRONTEX – facts and myths', by Ilkka Laitinen: 'Summing up I would like to remind that FRONTEX activities are supplementary to those undertaken by the Member States. FRONTEX doesn't have any monopole on border protection and is not omnipotent. It is a coordinator of the operational cooperation in which the Member States show their volition. If some of our critics think it is not enough they should fix their eyes on decision takers, as FRONTEX only executes its duties described in the Regulation 2007/2004'. In J. Rijpma, Building Borders The Regulatory Framework for the Management of the External Borders of the European Union, EUI PhD thesis. See also the Commission Communication on European Agencies – The Way Forward COM (2008)135, p. 7, where FRONTEX is classified as an agency in charge of operational activities: 'Agencies can be classified in different ways. One useful way is to try to look at the key functions they perform'.

45. See Article 6 and 7, Regulation (EC) No. 863/2007, cited above.

46. See JHA Council of 5–6 June 2008; Commission COM (2009) 262 p. 18; earlier: COM (2008) 67, p. 5.
47. The number of sea border joint operations is increasing every year, together with the number of participating states. See FRONTEX Press Pack, available at: www.frontex.europa.eu. This finds confirmation also in the annual budgets: in 2008, the 62 per cent of the total budget (31,1 MEUR out of 50,635,) has been devoted to sea border operations, whereas the second largest portion of the budget (13% of the budget) is represented by training (of border guards). See COWI Evaluation Report, 15 January 2009, available at the Frontex webpage, p. 25.
48. 'Based on their bilateral agreements with Spain, Senegal and Mauritania were also involved with their assets and staff. The main aim of this joint effort was to detect vessels setting off towards the Canary Islands and to divert them back to their point of departure thus reducing the number of lives lost at sea. During the course of the operation more than 3500 migrants were stopped from this dangerous endeavour close to African coast'. News Release 'Longest FRONTEX coordinated operation – HERA, the Canary Islands' of 19 December 2006, available at: http://www.frontex.europa.eu/newsroom/news_releases/art8.html. For figures, see the statistic published on Frontex's webpage. Accessed on 4 April 2011.
49. Tondini, M., '*Fishers of men?* The interception of migrants in the Mediterranean Sea and their forced return to Libya', *INEX Paper*, October 2010, available at: www.inexproject.eu, 16. Accessed on 4 April 2011.
50. In 2006 and 2007 operations, FRONTEX statements declare that about 3000 migrants were intercepted, one third within the operational area, two thirds outside it. In 2008 operations, FRONTEX declared that no migrant was diverted back or deterred; instead some 15 facilitators were arrested. FRONTEX statistics for NAUTILUS 2008 tell that 16,098 migrants arrived to Italy, and 2321 to Malta.
51. Klepp, S., 'A contested asylum system: The European Union between refugee protection and border control in the Mediterranean Sea', 12 *European Journal of Migration and Law* (2010), pp. 1, 16.
52. Klepp, S., 'A contested asylum system', op. cit., p. 17.
53. See Human Rights Watch Report 'Pushed Back, Pushed Around', available at: http://www.hrw.org/en/reports/2009/09/21/pushed-back-pushed-around-0, p. 37. Accessed on 4 April 2011.
54. Tondini, M., '*Fishers of men?*' op. cit., 16.
55. COWI report and Baldaccini, A., op. cit., 239.
56. See for example the reports of Statewatch, website Migrants At Sea, available at: http://migrantsatsea.wordpress.com, Human Rights Watch, the RefWorld tool of UNHCR, available at http://www.unhcr.org/cgi-bin/texis/vtx/refworld/rwmain. Accessed on 4 April 2011.
57. Trevisanut, S., 'The Principle of *Non-Refoulement* at Sea and the Effectiveness of Asylum Protection', in A. Von Bogdandy and R. Wolfrum (eds), *Max Planck Yearbook of United Nations Law*, vol. 12. (http://www.brill.nl/max-planck-yearbook-united-nations-law-volume-12-2008 Leiden-Boston-Tokio: Brill, 2008), pp. 205, 245; S. Trevisanut, 'Maritime border control and the protection of asylum seekers in the European Union', *Touro International Law Review*, vol. 12 (2009), pp. 157, 159.

58. Treaty of Friendship, Partnership and Cooperation with Libya, signed in Bengasi on 30 August 2008. For a comment see 'Il trattato Italia-Libia di amicizia, partenariato e cooperazione', dossier no. 108/2009, under the direction of N. Ronzitti, available at: http://www.iai.it/pdf/Oss_Transatlantico/108. pdf. Additional Technical-Operational Protocol of 4 February 2009, not publicly available. Accessed on 10 May 2011.
59. UNHCR Press Release: 'UNHRC deeply concerned over returns from Italy to Libya', 7 May 2009.
60. See *ex multis*, see the complete and accurate reconstruction by V. Moreno Lax, 'Seeking asylum in the Mediterranean: Against a fragmentary reading of EU Member States' Obligations Accruing at Sea, *International Journal of Refugee Law*, vol. 23 (2011), pp. 174–220.
61. See Article 263 of the TFEU: this provision states that the EUCJ will also review 'the legality of acts of bodies, offices or agencies of the Union intended to produce legal effects vis á vis third parties'. This will make possible for the EUCJ to review the legality of acts of FRONTEX.
62. The multiannual Hague Programme, adopted at the European Council of 4 and 5 November 2004.
63. M. Besters, F. W. A. Brom, 'Greedy' Information Technology: The Digitalization of the European Migration Policy, *European Journal of Migration and Law*, vol. 12 (2010), pp. 455–70, at 457 ff.
64. See FRONTEX Programme of Work 2011; Council Document 5691/11, 25.1.2011, p. 29.
65. Ibd., p. 48.
66. COM (2008)68, p. 3.
67. COM (2008)68.
68. COM (2008)68 and MEMO/08/86 from the Council.
69. COM (2010)673 final.
70. 'More than €50 million in EU funds from the European Security Research Programme of the Commission's FP7 has already been allocated to the adaptation of military surveillance techniques to Europe's borders'. From B. Hayes (NGO Statewatch) 'The robot armies at our borders', for the Economist's European Voice, 2.10.2010, available at http://www.europeanvoice.com/article/imported/the-robot-armies-at-our-borders/69598.aspx. Accessed on 4 April 2011. A more specific example is the consortium SEABILLA, coordinated by Selex (a Finmeccanica company), which was awarded a €10 million 'research' contract to develop an EU sea border surveillance system (the total project cost is €15.5 million, the EC contribution is €9.8 million). This has been launched in the framework of the FP7 programme, aiming at making the EU 'most dynamic competitive knowledge-based economy in the world'. The consortium 'SEABILLA', which was awarded the tender, includes a host of arms companies and defence contractors (BAE Systems, EADS, Thales, Sagem, Eurocopter, Telespazio, Alenia, TNO and others) and promises to: (1) define the architecture for cost-effective European sea border surveillance systems, integrating space, land, sea and air assets, including legacy systems; (2) apply advanced technological solutions to increase performances of surveillance functions; (3) develop and demonstrate significant improvements in detection, tracking, identification and automated behaviour analysis of all vessels, including hard to detect vessels, in open waters as well as close

to coast. According to the project synopsis, these surveillance systems will be used for: (a) fighting drug trafficking in the English Channel; (b) addressing illegal immigration in the South Mediterranean; (c) thwarting illicit activities in open-sea in the Atlantic waters from Canary Islands to the Azores; in coherence with the EU Integrated Maritime Policy, EUROSUR and Integrated Border Management, and in compliance with member states' sovereign prerogatives. Source: 'Research or procurement? Finmeccanica-Selex awarded €10 million EU sea border surveillance contract'. The information was retrieved at http://neoconopticon.wordpress.com/2010/09/22/research-or-procurement-finmeccanica-selex-awarded-e10-million-eu-sea-border-surveillance-contract/. Accessed on 4 April 2011.

71. The dominant view is actually the opposite. On this point we can find many different and divergent figures, which are difficult to assess. What is uncontested is that we talk about ranges of thousands. As it about human lives drowned at sea, we prefer not to provide any figure. The phenomenon is serious and it is worth all our attention, in spite of the nitty-gritty numbers involved.

72. NeoConOpticon, op. cit., p. 41.

73. Hayes, B., (NGO Statewatch), 'The robot armies at our borders', for the Economist's European Voice, 2 December 2010, available at: http://www.europeanvoice.com/article/imported/the-robot-armies-at-our-borders/69598.aspx. Accessed on 4 April 2011.

74. NeoConOpticon, op. cit., p. 37.

75. Hayes, B., 'The robot armies at our borders', op. cit. As to the position taken by the Obama administration in it, see the recent interview of UN Special Rapporteur P. Alston, available at http://www.democracynow.org/2010/4/1/drones.

76. Arias Fernandez, G., CIDOB report.

77. Ibid., 131.

78. Léonard, S., 'FRONTEX and the Securitization of Migrants through Practices', paper presented at the Migration Working Group Seminar, European University Institute, Florence, 9 February 2011.

79. See the Commission's communication COM (2008)68 final, p. 3, where the Commission discusses the objectives of EUROSUR.

80. See Agamben, G., 'Al di lá dei diritti dell'uomo', *Mezzi senza fine*, Turin 1996, pp. 20–30.

8

Regulating Public-Private Modalities of Legitimate Innovation: An *Ex Ante* Analysis Framework

Maurits P. T. Sanders

8.1 Introduction

This chapter is about safeguarding public interests by promoting innovation in the energy sector. By concluding climate accords at national and international level, the Dutch government has committed itself to the target of 20 per cent of energy consumption being provided by renewable energy by 2020. To achieve such sustainability ambitions the government is relying on technological innovations in the energy sector. To give these technological innovations an actual chance of success, it is initiating policy projects with private parties in partnerships. An example of one such initiative is the *Salland Green Gas project*. Taking the project as a basis, this chapter illustrates that public interests can have innovation as an object and that policy tools (as safeguarding mechanisms) must respond to the dynamism in innovation processes. We go on to present an *ex ante* analysis framework to be used for the selection of legitimate forms of public-private partnership (PPP). The analysis framework is presented as a phased plan in which an attempt is made to find a balance between the legal-administrative values of effectiveness and legitimacy.

8.2 Changing governance structures in the energy sector through the dynamic of public interests

There have been many developments in the energy sector in recent decades. One of the most striking changes is the splitting up of integrated energy companies, with network activities being separated from production and supply activities. These activities have then been placed in newly established organisations, which have been coordinated differently.[1] Firstly, network companies have been set up. These companies are responsible for the network activities and are regulated hierarchically within the circle of government (Kist et al., 2008). Secondly, production and supply companies have been set up. The activities of these organisations have been placed at a distance from the government, which means that the operation of the production and supply companies takes place in the 'free market'.[2]

The concept of public interest is important to understanding the energy sector reforms. In its report 'The safeguarding of public interest' the Dutch Scientific Council for Government Policy (WRR) distinguishes between three types of interest, that is: (i) individual, (ii) social and (iii) public interests (2000: 19).[3] This categorisation furthers the conceptual definition of public interest. First of all, the WRR defines social interests as interests in which protection for society as a whole is desirable (2000: 20). The WRR then gives substance to the concept of public interest by stating that a social interest becomes a public interest if the government is concerned about the protection of a social interest based on a belief that this interest will not otherwise be done full justice (2000: 20).[4]

The energy sector reforms are the result of political stands on public interests. A number of parties have an important position in this debate. First of all, the European Commission is important: it can actually be regarded as the great protagonist of the liberalisation wave in the energy sector. The Commission puts the interests of a properly functioning internal common energy market and consumer protection at the forefront of the policy choices underlying the European liberalisation directives (Kist et al., 2008: 13). In this market the consumer has freedom of choice. As a result the consumer is not dependent on just one provider (a monopolist) for energy. This means that the buyer of energy can make a comparison between different producers and suppliers and then select the provider that (best) meets its energy demand. The idea is that as a result of this confrontation of supply and demand the price mechanism steers buyer and producer towards an optimum transaction.

Despite the European impetus in the liberalisation of utility sectors, the degree of liberalisation and the approach to it differ from one member state to another (Wilkeshuis, 2010: 35). Implementation in fact counts among the powers of the member states (Wilkeshuis, 2010), which is where the positions of national authorities, such as the Minister of Economic Affairs and the Dutch parliament in the Netherlands, come into the picture. Initially, the thinking of the Dutch minister was that a properly functioning internal common energy market, with associated consumer protection, could be achieved by positioning the integrated energy companies in the market as a whole, it is that production, distribution and network management should be placed on the market together. According to the Dutch Minister of Economic Affairs, the realisation of this thinking would require a privatisation drive of the integrated energy companies after the liberalisation wave. This proposal, however, was resisted by the Dutch parliament. In the ensuing political debate the parliamentarians did not just attach importance to the public interests mentioned above. They raised the fact that security of supply and distribution network quality also had to be safeguarded. This was because members of parliament feared that these interests would come under pressure if the activities of the integrated energy companies were made entirely remote from government. The outcome of the political debate is that the public interests, mentioned above, are safeguarded by separating the energy networks, economically and legally, from the production and supply of energy.[5]

Simultaneously with the liberalisation of the energy sector and the resulting splitting up of integrated energy companies, the issue of climate change has moved higher up the political agenda. In recent decades society has made an increasingly emphatic appeal to the government to combat changes in the climate. This social interest has been picked up by the government and therefore become a public interest. The 'New energy for the climate' work programme expresses this transformation. The report illustrates that the government is accepting its responsibility in the field of climate change. The document sets out how the Netherlands will have one of the most efficient and clean energy supplies in the European Union by 2020 (VROM, 2007: 3). To achieve this ambition, central government has concluded administrative accords[6] on climate and energy with local authorities. Agreements have also been made with the business community, which are set out in so-called sustainability accords.[7] The policy initiatives for combating climate change are formulated in both the administrative accords and the sustainability accords. The following targets are key to the policy: (i) a 30 per cent cut in greenhouse gases by 2020 compared with 1990, (ii) an energy conservation percentage of 2 per cent per year, and (iii) renewable energy sources accounting for 20 per cent of energy generated by 2020 (VROM, 2007). To achieve these policy objectives, public and private parties are collaborating on innovative sustainability projects. An example of such an initiative is the Salland Green Gas project (see Box 8.1).

Box 8.1   Green Gas Project

On 14 January 2009 central government and the Association of Provincial Authorities (IPO) signed the Climate and Energy Accord between central government and provinces. By signing, the provinces were endorsing the importance of the national and European sustainability ambitions. The parties also agreed that in addition to their statutory task, the provinces would contribute to the achievement of climate objectives as they saw fit and if necessary with other public and private partners.

Even before the signing of the accord, the importance of sustainable development had been identified and recognised by the Province of Overijssel, as expressed, for instance, in the Overijssel Energy Pact Programme. With the energy pact the province was aiming to cut its CO_2 emissions by 30 per cent by 2020 compared with 1990 levels (Province of Overijssel, 2008). According to the province, this target can be achieved by an intensive and long-term approach, in which partnership arrangements are concluded with partners in the policy domain. This approach has two spearheads, that is: (i) nergy conservation in households and businesses, and (ii) the sustainable generation of energy. The plans for sustainable generation of energy have been formulated by the province in its Bioenergy sub-programme.

The Bioenergy sub-programme is a major component of the Energy Pact for the reduction of CO_2 emissions and will attempt to achieve 52 per cent of the total CO_2 reduction. To achieve this, the province is, for instance, making efforts to get the production and supply of biogas and/or green gas to consumers (households and/or businesses) off the ground. A critical success factor for the production and supply of both biogas and green gas is a regional energy infrastructure, known as a Green Gas Hub. A Green Gas Hub is an infrastructure for the production, supply and offtake of green gas. Producers of biogas, such as pig farmers and market gardeners, supply their biogas to a central plant through a pipe. In this central plant the biogas is upgraded to natural gas quality (green gas) and the end product can be fed into the natural gas network.

The ambition for a Green Gas Hub has been embedded in a concrete project by the Province of Overijssel. As the Green Gas Hub is being implemented in the Salland area, the project goes by the name of Salland Green Gas. Salland Green Gas can rightly be regarded as an innovative project. After all, there are only a few Green Gas Hubs in the Netherlands. This means that for both the technologies to be used and the administrative organisation, the project must be set up from scratch. To make this a success, the Province of Overijssel is working with the municipality of Olst-Wijhe, Raalte, Deventer, Saxion, Enexis and ROVA. Parallel to the execution of the project, the Province of Overijssel will in 2011 be elaborating the Green Gas Master Plan, which identifies the most promising clusters for bioenergy (potential green gas producers, large volume energy users, current infrastructure). The Salland Green Gas project is an integral part of this master plan.

The Salland Green Gas project is a concrete example of the efforts of the Province of Overijssel to combat climate change. The province has an active role in both the initiation and the execution of the project. This role-perception is in keeping with the role pattern for the government outlined in the 'New energy for the climate' work programme. The work programme talks of an initiating role for government so that test beds emerge for the application of innovative energy-saving techniques (VROM, 2007: 42). It is perfectly conceivable that such a role-perception is very much in keeping with policy initiatives to combat climate change. At the same time a fundamental question arises in the context of the earlier liberalisation of the energy sector. This did after all have the consequence of leaving the government remote from the energy sector. The question arises: in what way can public interest in a sector in which intensive government involvement is important be safeguarded? A solution to this problem appears to be available. The national 'New energy for the climate' work programme states in advance that there are opportunities for PPP.[8] Unfortunately, the Ministry of Housing, Spatial Planning & the Environment (VROM) omits any further elaboration of this thinking from the report, though it does appear in this chapter. It involves an assessment of the opportunities that actually exist for PPP below and the way in which a logical and consistent appraisal can be made for a form of PPP. This happens by presenting an *ex ante* analysis framework for legitimate forms of PPP. The focus is on legitimacy because it is generally accepted in public administration that public exercise of authority by the government must be legitimate and therefore also the regulation for this exercise of authority (cf. Weber, 1922; Luhmann, 1969; Beetham, 1991; Scharpf, 1998).[9] PPP is proposed by

VROM (2007) as a solution for realising innovations effectively, but this does not answer the legitimacy of PPP as control of innovation is not answered thereby and must therefore be examined on its merits. Below we first give an outline description of how legitimacy is approached in public administration.

8.3   Public administrative legitimacy

Legitimacy is one of the central concepts in public administration. In the literature this concept has therefore received considerable attention from various authors (see for instance: Weber, 1922; Luhmann, 1969; Beetham, 1991; Scharpf, 1998). For the presentation of an *ex ante analysis* framework for legitimate forms of PPP, David Beetham's perspective on legitimacy is important. In public administration, his body of thought can be construed as one of the most detailed perspectives on the legitimacy of public exercise of authority. His approach is described in outline below.

In the 1990s Beetham developed a cross-disciplinary perspective on legitimacy for the *ex post* evaluation of the legitimacy of public exercise of authority. In his judgment the analysis framework can be applied universally (1991: 21). To meet this ambition, Beetham regards legitimacy as a 'multi-dimensional' concept with three dimensions (1991: 15–16): (i) Legality, (ii) shared values and (iii) consent. Together they constitute the (cumulative) conditions for legitimacy, which can then also be tested empirically (Heldeweg and Sanders, 2011). Table 8.1 shows a summary of this:

Dimension 1., 'legality', requires the legally valid exercise of authority, as the 'law stands' – including unwritten rules (Beetham, 1991: 64 *et seq.*). Legal rules make (autonomous) social regulation possible,

Table 8.1   The three dimensions of legitimacy (Beetham 1991, p. 20, Table 1.1)

Criteria of Legitimacy	Form of Non-legitimate Power
1. Conformity to rules (legal validity)	Illegitimacy (breach of rules)
2. Justifiability to rules in terms of shared beliefs	Legitimacy deficit (discrepancy between rules and supporting beliefs, absence of shared beliefs)
3. Legitimation through expressed consent	De-legitimation (withdrawal of consent)

thanks in part to alignment of the government to the law – for which read, rule of law (Heldeweg and Sanders, 2011).

Dimension 2., 'shared values', requires that the rules by which we must act are also intrinsically justified (Beetham, 1991: 69 *et seq.*). They must be based on 'normative principles' that express shared value perceptions about the citizen-government relationship (standard setter and standard addressee respectively) and whose justification follows from their origin (what is the source?) or their content (why these rules?) (Heldeweg and Sanders, 2011).

Dimension 3., 'consent', refers to voluntary consent of the subordinate(s) with the political exercise of power by the dominant actor (Beetham, 1991: 91 *et seq.*), as by democratic mechanisms (Heldeweg and Sanders, 2011).

The author of this chapter takes the position that Beetham's analysis framework is not only applicable to the *ex post* assessment of legitimacy of public exercise of authority; his three-way structure can also be used as a starting point for the development of an *ex ante* analysis framework for legitimate forms of PPP in general and be applied to the promotion of innovation in particular. For this Beetham's analysis framework is adjusted and improved in a number of respects, which leads to a phased plan consisting of five steps, that is: (i) characterising of the interest, (ii) choice of the form of regulation, (iii) determination of the risks of failure, (iv) denoting PPP type and, finally, (v) determination of the legal form.

Step 1: characterising of the interest

Before the government concludes a PPP, it must first decide whether it has a role in the protection of an interest at all (see the distinction between types of interest made above in this chapter). As we have said, in the report 'The safeguarding of public interest' the WRR categorises three kinds of interest, that is (i) individual, (ii) social, and (iii) public interests (2000: 19). According to the WRR, in the case of social interests it is a matter of interests for which protection is desirable for society as a whole (2000: 20). The WRR says that the involvement of the government for the promotion of social interests is not necessary per se (2000). Many of these social interests are in fact protected without government involvement (2000: 20). But, according to the WRR, it is problematic to assume that the quality and the accessibility of given social interests, for instance, are adequately safeguarded without the involvement of the government (2000: 20). In those cases the government can concern itself with the protection of a social interest on the basis of the conviction that this interest will not otherwise be done full justice (WRR, 2000: 20). In such situations there is a transformation from a social interest into a public interest. This transformation means that the government makes the interest the object of its policy (WRR, 2000: 21). Which interests are public and in what way these interests are expressed in policy goals is the outcome of political argumentation and debate. Or, as the WRR puts it: in the case of public interests the 'what' question is pre-eminently a political one (2000: 21). The 'how' question (how the government must give shape to its final responsibility?) can be answered in many ways (WRR, 2000: 21). According to the WRR, in the case of the 'how' question, the key question is always whether the government must fulfil its final responsibility on its own or by bringing in private parties (2000: 21). The latter happens, for instance, with PPP. In such a partnership relationship between the government and private parties it is, however, possible to regulate the public interest in different ways. An assessment can be made between these forms of regulation. Step 2 is about this choice. The basic principle here is that innovation in the energy sector, in particular where the promotion of sustainable energy generation is concerned, is a public interest.

Step 2: choice of the form of coordination

Where the government has taken the final responsibility for a particular social interest and gives its preference for realising this interest in a PPP, the question about the way in which this partnership can be coordinated becomes topical. In public administration it is customary to make a distinction between three forms of coordination, that is: (i) market, (ii) hierarchy, and (iii) network (see Thompson 1991: 1).[10] These variants are first described in outline below, before a link with PPP is made.

The three forms of coordination are regarded in the literature as arenas in which goods and services (so policy too) come about through transactions between those concerned (public and/or private actors). This implies that the government can make a targeted choice of one of these forms of regulation in order to create a good or a service. Where the government formulates policy for the realisation of a public interest, it therefore has the choice of: (i) market, (i) hierarchical, or (iii) network coordination.

First of all there is the market. This type of coordination is not in itself a priori targeted. The government can however use market regulation for the realisation of a public interest, in particular to secure efficiency. Efficiency is not then the goal, but the outcome. The basic thinking behind such an outcome is that both 'demanders' and 'providers' of a good or service participate in transactions willingly. Together these economic actors form an arena, in which the transactions take place. The idea is that the actors make an assessment between all the conceivable alternatives on the basis of full information and in their choice pursue the optimisation of their own welfare function.

Hierarchical coordination differs from market coordination in the sense that a hierarchy is in itself targeted. In this case an attempt is made to realise a public interest by government control. The parties participating in this process are expected to act 'in the spirit' of the objective laid down or to submit to it according to given rules of behaviour. For the optimisation of this process it is, for instance, necessary to define tasks precisely, to grant few autonomous powers to contract partners and to structure responsibility relationships top-down. In contrast to market control, in the hierarchy efficiency is far more a by-product of the control, which itself is primarily directed at the effective protection of public interests.

The third arena in which goods and services come about is the network. Just as in the case of the hierarchy, a network is set up targeted to pursue policy. The network is to be perceived as a setting in which the participating autonomous actors try to reach agreement on a strategy to achieve a policy goal. It is not the price mechanism that controls the transactions in this case, as is the case with the market, nor unilaterally one particular authority, as in a hierarchical setting; it is the interaction between the participating parties that leads to agreement on the policy strategy. This interaction is necessary because the actors have an equal relationship. This equality arises through the specific knowledge or expertise that the parties contribute. Parties in a network are equal on the basis of mutual dependence.

As we have indicated, for the safeguarding of a public interest an assessment must be made between the three arenas. Guiding in this choice process are the value orientations underlying the forms of coordination. These value orientations differ from one type of coordination to another.

The basic thinking behind market forces is that economic agents acting rationally on the basis of full information (the price of an economic good) make an assessment leading to an optimum outcome (efficiency). This outcome is, however, only possible where there is a perfect market (in other words, there can be no question of market failure). Perfect market forces exist where the market satisfies two basic principles, that is: (i) it must be 'open' (transparent), and (ii) there must be fair competition. In the practice of public administration these two principles are, for instance, expressed in the European tendering directives.

Table 8.2   Underlying value orientations of forms of coordination

Step 2: Choice of the form of coordination

	Market	Hierarchy	Network
Underlying value orientations	Transparency and fair competition	Democracy, liberal rule of law and servient government	Reciprocity and general acceptance

In hierarchical coordination it is not so much a matter of competition, but of exercise of public authority by the government (or by executives in companies, but this interpretation of hierarchical control is not under discussion here). For the exercise of public authority in the Dutch doctrine a number of leading legal political values are maintained, which have been clustered below in three dimensions, that is (Zijlstra, 2009: 6–8; Heldeweg and Sanders, 2011): (i) 'democracy': voice of citizens over government power (with the primacy of general people's representation – representation over participation); subsidiarity and decentralisation; openness, (ii) 'liberal rule of law': the separation of state and society, the primacy of civil autonomy and alignment of the government to the law – in particular spreading government authority, legality, fundamental rights, legal protection (and embedding in the international legal order), (iii) 'servient government': the government does not exist for itself but for social justice and should achieve this effectively and efficiently.

Networks are described in public administration as more or less stable patterns of relationships between mutually dependent actors formed around policy issues or policy programmes (Kickert, Klijn and Koppenjan, 1997: 6). In such an arena policy is the outcome of effective communication. For this communication in the first place the basic principle applies that there must be general acceptance. This means that all the interests in the network must be represented and that the policy outcome is then accepted by all the participants. The second basic principle is that there must be reciprocity. This means that the participants act without a direct consideration being provided.

Table 8.2 shows a summary of the underlying value orientations of the forms of coordination.

Step 3: determination of the risks of failure

It is important to realise that any form of coordination in innovative policy projects involves the risk of a given type of failure. The step

Table 8.3   Forms of failure

Step 3: Determination of the risks of failure

	Market	Hierarchy	Network
Forms of failure	Missing investments, negative external or distribution effects	Supply control and bureaucratic inefficiency	Inclusivity/ exclusivity, hierarchy within the network

following the choice of an arena to bring about an innovative policy project is therefore the indication of these risks. A worthwhile distinction that is in line with the different forms of regulation is: (i) market failure, (ii) government failure, and (iii) network failure. In the practice of public administration these categories of failure in innovative policy projects can have all kinds of forms of expression. Given this diversity, preference has been given to deepening these forms of expression, beyond the scope of the chapter. Table 8.3 gives a number of examples of each type of failure.

Step 4: denoting PPP type

It is important to indicate the failure factors in the previous step because they can be avoided by introducing subsets in the three arenas that consist of elements of the other forms of regulation. The consequence of combining elements of different forms of regulation is that hybrids emerge. PPP must be understood as an example of such a hybrid.[11] PPP is defined by Michiel Heldeweg and Maurits Sanders as a legally structured partnership between one or more authorities and one or more corporate entities governed by private law that focuses on the development and execution of a common strategy for the realisation of a policy project (2011).[12] The above implies that a PPP is possible in the three alternative forms of coordination. For this reason Heldeweg and Sanders distinguish the following types of PPP: (i) 'PPP in a market arena' or 'market PPP', (ii) 'PPP in a network arena' or 'network PPP', and (iii) 'PPP in a hierarchical arena' or 'authoritative PPP'. Table 8.4 gives a summary of the forms of PPP.

The goal of the market PPP is not this joint strategy, but putting a policy project into effect on the basis of (mutually beneficial) exchange, against the background of separate public and private positions (Smit, 2010). This leads to PPP as bare configuration, as in the construction of durable property such as wind farms (Heldeweg and Sanders, 2011). The

Table 8.4   Forms of PPP (Heldeweg and Sanders 2011, Table 1)

PPP type	Goal and Approach (way of working)

Market PPP	Goal: to put a policy project into *effect*
	Approach: as *exchange*, on the basis of *separated* powers and responsibilities
	   – public party lays down – in particular assignment and decisions
	   – private party puts into effect – in particular 'working & services'

Network PPP	Goal: *joint determination* of goals and an associated strategy as regards a policy project
	Approach: focused on *coordination* of powers and responsibilities
	   – public party lays down
	   – private party may participate in the putting into effect

Authoritative PPP	Goal: *authoritative determination* (and arranging putting into effect) of a policy project
	Approach: on the basis of *joint* powers and responsibilities
	   – joint determination of goals/strategy/decisions (with public authority)
	   – putting into effect itself or by others

government formulates the project and then the phases, from building design and building construction through to maintenance and/or operation, are put out to tender, for example, in the form of a Design Build Finance Maintain (Operate) (DBFM(O)) contract (van Ham and Koppenjan, 2002).[13]

In the 'network PPP' there is partnership in an association (let's say, a 'committee') with the aim of formulating a joint strategy – such as a municipality consulting with retailers about the design of a town centre (Heldeweg and Sanders, 2011). According to Heldeweg and Sanders, it does, however, remain the case that public and private parties have their own tasks, powers and responsibilities, so that implementation of the strategy by formal decisions to this effect remains a matter for the government (and it otherwise remains to be seen – possibly in terms of competition law – whether and if so what role the private parties concerned play in any execution) (2011).

Finally, Heldeweg and Sanders distinguish the 'authoritative PPP', which not only has partnership at strategy formulation level, but its determination or the taking of (execution) decisions to this effect also counts among the powers of the partnership; this PPP goes hand

Table 8.5   Summary of legal forms[14]

Step 5: Determination of the legal form

	Market	Hierarchy	Network
Legal forms	Only contract, joint companies (partnership, general partnership, limited partnership, public company, private company, foundation and/or (cooperative) association)), property law arrangements (ownership, easements, qualitative obligations, perpetual clauses, ground lease, building and planting rights, apartment rights)	Administrative committees, provincial committees, joint regulation	Optional committees

in hand with citizen-binding decisions, in short: public authority (Heldeweg and Sanders, 2011).

Step 5: determination of the legal form

Once it has been determined at Step 5 which form of PPP is involved, a concrete form of organisation can then be laid down. Different arguments play a part in the assessment process for a given legal form, which also differ from one situation to another. The participants in a legal form may, for example, attach importance to limitations of liability or to tax transparency. In addition to participants, restrictions are also laid down by the authorities – as outcomes of multi-level governance, for example, through EU rules. By identifying these arguments and frameworks it is possible to make the definitive choice for a legal form. Table 8.5 gives a summary of legal forms.

8.4   Conclusion

On the basis of developments in the field of sustainability in the energy sector, this chapter has illustrated that what is regarded as public interest is the outcome of political argument and debate. This dynamic does not just have an impact on the interests that are regarded as public (the 'what' question), but also on the way in which public interests are safeguarded (the 'how' question). Different mechanisms are possible for the safeguarding of public interests.

In the field of renewable energy sources (as in the case of the Salland Green Gas project) the government is relying on technological innovations by private parties in the sector. The government sees it as its responsibility to offer these private parties the space for technological innovations. This has been made possible by concluding PPP links with parties in the policy domain.

This chapter shows that different types of PPP are possible for the safeguarding of public interests and that a variety of legal forms go with these types. This variety raises the question of how a legitimate form of PPP can be selected in a logical and consistent way. In this chapter an *ex ante* analysis framework has been presented for this. This analysis framework has been presented as the following phased plan:

1. characterising of the interest;
2. choice of the form of regulation;
3. determination of the risks of failure;
4. denoting PPP type;
5. determination of the legal form.

The above analysis framework is being applied in the Salland Green Gas project. Because this process is currently ongoing, it is not yet possible to report on the outcome of the *ex ante* analysis framework.

Notes

1. Thompson distinguishes three forms of regulation, that is: (i) market, (ii) hierarchy, and (iii) network (see Thompson et al., 1991).
2. The shares of these companies are sometimes in government hands and sometimes not (any more).
3. The concept of public interest is studied in various scientific disciplines, which has led to different definitions of the term. Van Genugten categorises these different meanings in two types of approach, that is: (i) economic approach to public interests, and (ii) public interests in the politico-administrative reality approach (2008, p. 5). Because of the administrative-legal approach of this chapter and the casuistry, a definition of public interests from the politico-administrative approach has been chosen. The WRR approach falls into the politico-administrative approach (Van Genugten 2008, p. 5). Alternative definitions can also be found in the politico-administrative approach, which usually cover the same overtone. An example of this is the Socio-Economic Council (SER) which, in the design advice 'Public interest requires customisation in market mechanism', describes this type of interest as interests whose protection is desirable for society as a whole and that politics is concerned about for this reason (2010). In the opinion of the author, this description is an excellent conceptual starting point in this chapter in view of the authoritativeness of the WRR definition.
4. For the realisation of public interests the government then formulates policy. Because the term 'policy' is used regularly in this chapter, a conceptual

definition is important. A topical and frequently quoted definition is given by Hoppe et al (2004). His definition is 'policy is a politically confirmed plan for the approach, preferably a solution, to a social problem' (Hoppe et al., 2004, p. 14).

5. See the Independent Grid Management Act.
6. The climate accord 'Working together on a climate-proof and sustainable Netherlands', which was signed by central government and the Association of Dutch Municipalities (VNG) on 12 November 2007, is an example of such an administrative accord. These climate accords also exist between the Association of Regional Water Authorities and central government and the provinces and central government.
7. An example of a sustainability accord is the accord signed by the Cabinet and the business community (MKB-Nederland, VNO-NCW and LTO Nederland) on 1 November 2007.
8. The European Commission is also seeking its salvation in PPP, witness some major initiatives, especially where innovation is important. An example of this is the Green Cars initiative.
9. Discussion of this basic principle falls outside the scope of this chapter.
10. The forms of coordination can be approached in two ways. First, as spontaneous coordination, in which no direction is given, but emerges in the context. For example, a group of people having intensive dealings with each other can have the characteristics of a network. Second, the forms of coordination can be approached as chosen coordination. This means that before the production of a good or a service an arena is chosen. For example, the government has the choice of leaving policy projects to the market, organising them hierarchically or letting them come about in a network context. In this chapter the forms of coordination are approached as chosen forms of coordination.
11. Committees for interactive policy forming are another example of a hybrid form.
12. The Heldeweg and Sanders definition is in part based on the definition used by Bregman (2005).
13. Design, Build, Finance, Maintain (and Operate) (Van Ham and Koppenjan, 2002) – with interaction (cf. the 'competition-oriented dialogue'), but not as independent goal (as with network and authority PPP).
14. This summary is partly based on Bregman (2005).

References

Beetham, D., *The Legitimation of Power* (New York: Palgrave Macmillan, 1991).
Bregman, A.G., *Publiek-Private Samenwerking Bij de Ruimtelijke Inrichting en Haar Exploitatie* (Bouwrecht Monografie nr. 26, Deventer: Kluwer, 2005).
Genugten, M. van, *The Art of Alignment: Transaction Cost Economics and the Provision of Public Services at the Local Level* (Enschede: Proefschrift Universiteit Twente, 2008).
Ham, H. van and J. Koppenjan, *Publiek-private samenwerking bij transportinfrastructuur. Wenkend of wijkend perspectief?* (Utrecht: Lemma, 2002).
Heldeweg, M. A. and M. P. T. Sanders, *Botsende publieke waarden bij publiek-private samenwerking. dimensies en dilemma's van juridisch-bestuurskundige legitimiteit, in*

het bijzonder bij openbaar gezag, Tijdschrift voor Bestuurskunde, Den Haag: Reed Business, (20) 2, pp. 33–43, 2011.

Interprovinciaal Overleg (IPO), *Klimaat- Energieakkoord tussen Rijk en provincies*, Den Haag, 2009.

Kickert, W. J. M., E. H. Klijn and J. F. M. Koppenjan (eds), *Managing Complex Networks: Strategies for the Public Sector* (London: Sage Publications, 1997).

Kist, A. W., F. J. M. Crone, D. F. Hudig, N. G. Ketting, T. de Swaan and R. Willems, *Publiek aandeelhouderschap energiebedrijven*, Ministerie van Economische Zaken, Den Haag, 2008.

Luhmann, N., *Legitimation durch verfahren* (Frankfurt am Main: Suhrkamp, 1969).

Provincie Overijssel, *Uitwerking Programma Energiepact Overijssel. Statenvoorstel nr. PS/2008/375*, Zwolle, 2008.

Scharpf, F. W., *Games Real Actors Play: Actor centered institutionalism in policy research* (Oxford: Westview Press, 1998).

Smit, M., *Publiek belang: hoe houd je het op de rails. Een studie naar de effectiviteit en legitimiteit van planvorming voor stationslocaties* (Enschede: Proefschrift Universiteit Twente, 2010).

Sociaal Economische Raad (SER), *Publiek belang vraagt maatwerk in marktwerking*, Den Haag, 2010.

Thompson, G., J. Frances, R. Levacic and J. Mitchell, (eds), *Markets, Hierarchies & Networks: The Coordination of Social Life* (London: Sage, 1991).

Vereniging Nederlandse Gemeenten (VNG*), Samen werken aan een klimaatbestendig en duurzaam Nederland. Klimaatakkoord Gemeenten en Rijk 2007–2011*, Den Haag, 2007.

VROM, *Nieuwe energie voor het klimaat: Werkprogramma schoon en zuinig*, Den Haag, 2007.

Weber, M., *Wirtschaft und Gesellschaft* (Tübingen: Mohr, 1922).

Wetenschappelijke Raad voor het Regeringsbeleid (WRR), *Het borgen van het publiek belang*. Rapporten aan de regering nr. 56, 2000.

Wilkeshuis, K., *Publieke belangen en nutssectoren. Op weg naar een juridisch afwegingskader.* (Amsterdam: Proefschrift Vrije Universiteit, 2010).

Zijlstra, S. E., Bestuurlijk Organisatierecht (Deventer: Kluwer, 2009).

Part III
Emerging Technologies and Technology Innovation: Regulatory Issues, Actors and Patents

9
Transnational Regulation of Nanotechnology: Institutional Diversity in Agenda Setting and State Support

Nupur Chowdhury

9.1 Introduction

Risk regulation in the context of new technologies like biotechnology, gene therapy and nanotechnology has been one of the most contentious subject areas and has not only spawned reams of academic literature but has also driven investment in fundamental research to investigate such risk perceptions (Fraiberg and Trebilcock, 1998). The risk debate at the international level emerged primarily in the context of international trade regimes. Prior to the debate on GM (genetic modification) technology, risks emanating from specific technologies were seen to be primarily a technocratic issue which would be dealt with domestically by national governments (Jasanoff, 1995; Hackett et al, 2008). With the establishment of the WTO, a number of issues which were hitherto considered to be primarily domestic policy issues were linked to the international trade regime and consequently became legitimate areas of international policy making and regulation. IP rights, food standards and environmental standards all became linked in terms of their impact on international trade (Berstein and Hannah, 2008; Jones, 2002) and therefore their regulation would have to conform to certain agreed norms of equity and proportionality underlying the international trade obligations of member countries. However the past decade has witnessed a number of transnational regulatory schemes that have been promoted by a range of actors – including intergovernmental organisations, NGOs and other private actors – that have been developed outside the traditional normative framework of ISO and the Codex Alimentarius (Abbott and Snidal, 2010).

The explosion of academic literature within the area of regulatory governance of emerging technologies or technologies, which had certain associated environmental and health risks associated with it, reached a fevered pitch during the height of the controversy surrounding GM (Cantley, 1995; Barling, 1997; Chambers, 1999; Buonanno, Zablotney and Keefer, 2001). The advent of nanotechnology has also witnessed increasing volumes of academic debates and literature being produced on similar questions about the regulatory options that should be explored in the context of the uncertain health and safety risks that are associated with manufactured nanomaterials and therefore the nano applications based on them (Fielder and Reynolds, 1994; Mehta, 2002; Bowman and Calster, 2007; Lin, 2007; Bowman and Hodge, 2008; Bowman and Calster, 2009; Dorbeck-Jung, Bowman and Calster, 2011). There are subtle differences between the two seemingly similar debates on these technologies, which hold the promise of drastic transformation of this twenty first century. First, the public fear vis-à-vis biotechnology was basically focused on the agriculture sectors, specifically with reference to GM technology. Nanotechnology or manufactured nanomaterials, on the other hand, have a range of applications across disciplines and production sectors. Nanotechnology poses a far greater challenge in terms of identifying specific characteristics and also applications that could be a regulatory challenge on the basis of uncertain risks. The applications being of a dispersed nature also preclude alliance formation and civil society activism with a clear focus and policy objectives. Given the wide number of applications associated with nanotechnology it becomes difficult to delimit and focus efforts on a particular product sector. This partially explains the limited extent of lobbying by national NGOs on issues associated with nanotechnology governance as compared to biotechnology. Most of the activism from civil society organisations has been spearheaded by international NGOs like Action on Erosion, Technology and Concentration (ETC) Group and Friends of the Earth. Second, concentrated efforts have been made by the public administrations in most countries to differentiate between these two technologies and, to a large extent, push the argument that the current state of scientific knowledge is still insufficient and therefore more research investment is required.[1] This has been used to delay any new regulatory actions in this field. Thus nanotechnology development being still at its nascent stage, regulatory agencies face an information deficit and this is seen to be an impediment in choosing regulatory options (Wilson, 2006). Although widely stated, this is a half-truth. There are already more than 800 nanotechnology-based applications that have been launched in the market.[2] This number is expected to grow rapidly in the short and medium-term (Felcher, 2008).

These developments seem to suggest that most regulatory agencies have acknowledged the need to strengthen international efforts for cooperating on standards, risk regulation, etc. This may seem to make this debate irrelevant. Nonetheless revisiting the debate is important because it may have some impact in defining the boundaries of international regulation vis-à-vis domestic regulation. Gary E. Marchant and Doughlas J. Sylvester (2006) contend that various rationales exist that reinforce the need for international regulation of nanotechnology. These include the risk aspect of nanomaterials and the potential for transboundary harm, the uncertainty surrounding the behaviour and health and environmental impacts of nanomaterials and therefore the need to share regulatory expertise and resources. The aim of preventing trade protectionism by international harmonisation of risk regulation and standards is a perennial one that has been used earlier in the case of biotechnology. Reducing the inequities of technology access and development in terms of a north-south debate has also been echoed in the context of nanotechnology and remains to be a contentious issue within international policy making. One could make the argument that these rationales exist for other emerging technologies too and are therefore not specific to nanotechnology. To a certain extent that is correct. Similar arguments were also made in the context of biotechnology (Francioni , 2006). However in one aspect, there is a clear difference in degree. Nanotechnology represents a far greater challenge in terms of standardisation and evaluating potential environmental and health of nanomaterials, partly because of the nature of nanotechnology itself, which is a platform technology and in which applications are therefore bound to figure in a variety of modes and sectors. In this context the incentive for sharing resources in investigating these various aspects is much greater, given that the cost of investment is much greater. Regulatory partnerships therefore make sense not only because of efficiency arguments but also in terms of ensuring influence and securing strategic interests of the nations.

Internationally, several initiatives launched in the last few years have focused on an entire range of activities from those of pre-regulation, like research on values and standards for manufactured nanomaterials, to exploring regulatory options like the application of the precautionary principle. Apart from the multiplicities of such activities along the regulatory continuum, another characteristic feature of these developments is the involvement of multi-level actors. They include national actors such as DEFRA (Department for Environment, Food and Rural Affairs) in UK, the US EPA (Environment Protection Agency) and civil society organisations like Environment Defense, Friends of the Earth (Australia), regional actors like that of the EU, ASEAN (Association of South-East Asian Nations) and APEC (Asia-Pacific Economic Cooperation) and lastly international actors, viz. ISO, OECD, IRGC, IFCS (Intergovernmental Forum on Chemical Safety – an expert group promoted by the World Health Organisation (WHO)) and the ETC Group. As is apparent from this listing (not exhaustive), the actors involved consist of both public and private organisations.

The focus herein will be on three such agencies, the OECD, the IFCS and the IRGC. Apart from reasons of economy, there are several arguments that underlie the choice of these three actors. First, all these actors have displayed tremendous interest and have developed concrete plans to undertake policy development vis-à-vis nanotechnology. This has enabled them to take a lead in positioning themselves as reference points for future international policy making on nanotechnology. Second, interestingly the basis for involvement of these three actors with nanotechnology, have been largely an internal mandate.[3] Third, these actors have made some progress in developing mandates and carving out specific areas of influence and therefore could be studied to understand whether the political dynamics of the sector influences their capability and functional spaces. Fourth, all the agencies selected represent different organisational setups; that is, the IRGC is a largely non-governmental grouping of experts, the IFCS is an international grouping of experts within the intergovernmental regime of the WHO. The OECD, on the other hand, started out as a regional entity, which has expanded its membership to include economically stronger nations across other regions. These differences in origin, membership and structure, provide interesting insights illustrating the influence of such structural factors on developing policy competencies in newer areas, like, in this case, nanotechnology.

The primary aim of this paper is to provide a descriptive overview of the nanotechnology regulatory activities undertaken by the OECD, IFCS and IRGC. Using the theoretical framework of governance triangle developed by Kenneth W. Abbott and Duncan Snidal (2010), I identify the role played by each of these actors in the agenda-setting phase of the regulatory process. It should be underlined that agenda-setting assumes critical importance in the context of nanotechnology – especially because of the uncertain health and safety risks, and therefore the need to adopt transparent and representative processes to engender public trust. Further, fundamental public interest choices may be made that will limit the public policy options that may be available for negotiation at a later stage. I explore whether each of these actors individually have the competences to act effectively in the agenda-setting stage and to what extent they have tried to incorporate competences to fulfil any competency deficits. My focus here is also on the background role played by states in supporting certain actors over others and thereby leading to their emergence as the pre-eminent actors in this domain. Ultimately the question is also raised whether privileging certain kind of actors over others and thereby limiting access to agenda setting activities will mean that the standards resulting from this process will be able to fulfil common interest objectives (Mattli and Woods, 2010)?

9.2 Institutional diversity in international actors: IFCS, OECD and IRGC

The IFCS describes itself as a 'transparent and inclusive forum for discussing issues of common interest and also new and emerging issues in the area of sound management of chemicals'.[4] It is therefore a forum for experts and other interested parties to discuss and debate management issues relating to chemical safety (Gartner, Kullmer and Schlottmann, 2003). Another interesting aspect is that the forum was conceived at the UN Earth Summit at Rio de Janeiro in 1992, generally as a non-institutional arrangement for enabling governmental and non-governmental organisations to meet and discuss wide-ranging issues dealing with chemical safety and environmentally sound management of chemicals. The WHO serves as the administering authority and provides the secretariat for the IFCS. This umbilical linkage to the UN system provides considerable legitimacy and is the basis for the authority of the decisions that it adopts in its various sessions (known as forums). The IFCS is an example of a PPP operating at a global level. Essentially the drive has been to promote a consensual decision-making framework based on discussions and debates between the different members. Its decisions are non-binding in nature but carry substantial weight in establishing an international normative standard and form an important part of international soft law on chemical management.

The OECD is an international organisation, with membership made up of 30 countries primarily drawn from Europe but also including the USA, Australia and Japan. In the recent past, it has made overtures to expand membership to emerging economic powers like Russia, Brazil, China and South Africa.[5] The focus of its activities lie in economic policy areas and trade. The aim is to provide a forum for discussing comparative economic problems and policy prescriptions both internationally and across its member countries. It is an intergovernmental organisation focusing on a wide range of economic policy issues. Its activities are consultative in nature, however, given its membership and strong focus on economic research and analysis, policy prescriptions emanating from it have a substantial impact in identifying and pushing for policy choices.

Of the three actors, the IRGC is the most 'private' in nature. By this I mean that there is no overt organisation linkage between it and other international or national organisations that have a formal mandate to undertake work on global issues. By its own admission, the IRGC's work includes developing concepts of risk governance, anticipating major risk issues and providing risk governance policy recommendations for key decision makers.[6] Despite any organisational linkages to the formal international policy making process, the definition suggests that the IRGC sees itself as an advisorial authority for policy makers worldwide. The IRGC focuses on risk governance as an issue across sectors, viz. nanotechnology, climate change, disaster management and synthetic biology (among others). The aim is not only to contribute to the development of risk governance as a policy tool in general but also to focus on emerging global risks that require cooperative action across countries and both state and non-state actors. Its membership includes Swiss scientists, erstwhile policy makers and international, risk experts. It also aims at collaborating with a host of intergovernmental organisation, civil society organisations and national policy makers on issues of common interest. The twin aims of collaborating with such external persons is to not only draw legitimacy for their work (directly through peer review but also indirectly through claiming interest representation) but also help to create channels of influence through the policy recommendations could be circulated and adopted by formal state actors.

9.3 Governance triangle in the context of nanotechnology

Abott and Snidal's theoretical framework of a governance triangle was developed in the context of transnational regulatory standard-setting activities. They identified the state, firms and NGOs as three actor types that have engaged individually or collaboratively in regulatory standard-setting at the transnational level. The regulatory process has been divided into distinct phases of standard-setting – agenda-setting, negotiations, implementation, monitoring and enforcement (ANIME). Their basic premise is that in order to succeed in the regulatory process – actors will require four basic competences – viz. independence, representativeness, expertise and operational capacity. In most cases, actors acting singly will not have all the necessary competences and thus, collaboration is essential if they are to act effectively in the regulatory process (Abbott and Snidal, 2010). They use this framework to map transnational regulatory standard-setting schemes – into pre and post-1985 schemes.

This set of examples suggests that there is great institutional diversity among the schemes that actors have been involved in. The pattern of regulatory governance has also shifted away from single actor schemes to more collaborative and multi-actor schemes. Abott and Snidal also emphasise the background role played by the state in shifting the bargaining balance between firms and NGOs in facilitating the achievement of socially desirable goals. However, they also admit that the state's motivations are equally diverse, depending on both internal factors (the level of economic development) as well as external factors such as trade interests. This is evident from their support of intergovernmental organisations as preferred forums of transnational regulatory standard-setting (Abbott and Snidal, 1998).

Given this context, all the three actors discussed here – OECD, IFCS and IRGC – can be characterised as representative of the archetypes as discussed in the context of governance triangle. OECD is an intergovernmental organisation that is dominated by states. IFCS is a collaborative scheme formed of states and NGOs, and the IRGC can be typified as a firm – or a private actor – acting in its own self-interest. Currently transnational nanotechnology regulation can be characterised as being at the agenda-setting stage (although it is poised to move towards the negotiation stage).[7] It needs to be underlined that the agenda-setting stage is especially critical in the context of nanotechnology – given the health and environmental safety concerns that are associated with it. At this stage, the theoretical framework identifies states and NGOs as relatively important actors given that their competencies match the requirements of the agenda-setting stage. In the following section I provide an account of the nanotechnology-related regulatory activities undertaken by the three actors.

9.4 Nanotechnology regulatory action by international actors

Intergovernmental forum on chemicals safety

In 1994 the International Conference on Chemicals Safety established the IFCS and constituted its first meeting in Stockholm. The aim was to constitute a mechanism through which governments, intergovernmental organisations and NGOs could work together to promote chemical safety and the environmentally sound management of chemicals globally. Since then the IFCS has had six meetings, the last being IFCS Forum VI at Dakar, Senegal in September 2008. One of the final outcomes of this last meeting was the Dakar Statement on Manufactured Nanomaterials.[8] The negotiations focused on the role and the nature of the engagement that IFCS should have in the context of developments in nanotechnology and also on the possible policy options available in ensuring chemical safety. Another important aspect of the IFCS VI negotiations was on its future, with the development of the Strategic Approach to International Chemicals Management (SAICM) under the ICCM (International Conference on Chemicals Management). This is an important aspect since, as we shall see, the decision of the ICMM2 not to accept the IFCS offer of integrating it as a subsidiary body to ICCM[9] may affect its very existence and therefore its capability to engage with safety issues in nanotechnology globally. It reflects a deliberate choice on the part of Europe, the US and other leading countries that are exponents of nanotechnology, to support certain intergovernmental forums with a clear division cf policy making between them in pushing for international harmonisation and policy convergence on standardisation, environmental and health safety and IP rights in nanotechnology.

The primary issue was whether the IFCS mandate being limited to chemical safety, could also include social and ethical implications of nanotechnology. Several NGOs and developing country participants supported its inclusion within the IFCS agenda. However most European countries underlined the distinction between nanotechnology and manufactured nanomaterials, and supported the inclusion of the latter (as far as chemical safety aspects were concerned) into the IFCS agenda. This was resolved in the Dakar statement, by including a preambular reference 'to the need to address safety aspects of nanotechnology' but choosing to limit the focus of the statement only to nanomaterials. Product labelling of those containing nanomaterials and the need for a global code of conduct was also a point of heated debate. The final statement reflects these divisions in its considerable watering down to only specifying that the feasibility of global codes of conduct needs to be evaluated and that information requirements on manufactured nanomaterials could be met through product labelling, but also websites and databases. Nevertheless the statement was the first policy recommendation given by any such international forum on the issue of safety of manufactured nanomaterials.

The Dakar Statement on Manufactured Nanomaterials recommended applying the precautionary principle as a general principle of risk management both by the government and the industry during the entire lifecycle of the manufactured nanomaterials. Access to information was also sought to be facilitated by cooperative actions across and between actors; viz. manufacturers, researchers, governments and other stakeholders. Partnerships should also be established for channelling support to developing countries and economies in transition in building scientific, legal, regulatory and technical expertise on the issue of risks associated with manufactured nanomaterials. Finally it called upon the ICCM2 to consider these recommendations for further actions.

Organisation for Economic Cooperation and Development

There are essentially two mechanisms through which the OECD has undertaken policy making on nanotechnology. The OECD's Working Party on Manufactured Nanomaterials (WPMN), was established in 2006. It is a subsidiary body under the Chemicals Committee and focuses on the human health and environmental safety aspects of manufactured nanomaterials. The other is the Working Party on Nanotechnology (WPN), which was established by the Committee on Scientific and Technology Policy (CSTP) in 2007 and serves as a forum for discussions on emerging policy issues from nanotechnology developments globally. The 2009–10 work programme of the WPN focuses on developing statistical methodologies in order to benchmark and compare developments in nanotechnologies, addressing challenges to the business environment specific to nanotechnology, fostering international scientific cooperation in nanotechnology and in selecting key policy issues (viz. risk governance), and enabling nanotechnology while responding to specific global challenges like climate change, energy security, water, health and the environment. Peripherally, the OECD has also cooperated with Allianz, in bringing out a report on 'Opportunities and Risks of Nanotechnologies' under its International Futures Programme.

The WPMN has put together a group of eight projects that focus on generating information on health and environmental safety aspects of manufactured nanomaterials and on developing internationally harmonised hazard exposure and risk assessment standards. The projects include the following: first, development of a database on human health and environmental safety research; second, research strategies on manufactured nanomaterials; thirdsafety testing of this representative set of nanomaterials, and developing test guidelines; fourth, cooperation on voluntary schemes and regulatory programmes; and fifth, on the role of alternative methods in nanotoxicology and exposure measurement and exposure mitigation. The fifth project mapped the national information-gathering programmes (whether voluntary or mandatory) and has identified similarities and differences between such programmes. It also prepared recommendations on mechanisms and elements while setting up information-gathering initiatives. The steering groups of the third and the fourth projects are co-chaired by the United States and the European Commission. Under the third project a sponsorship programme has been launched for the testing of 14 specific manufactured nanomaterials. Several member states of the European Union like France, Germany, United Kingdom, as well as the European Commission itself, have become lead sponsors in the case of specific nanomaterials. In terms of private participation, it is the BIAC (Business and Industry Advisory Committee) representing major business organisations in the OECD member countries. Most of the publications of the WPMN have been developed and referenced with the Inter-Organisation Programme for the Sound Management of Chemicals (IOMC), thus ensuring their dissemination among the FAO (Food and Agriculture Organization of the United Nations), ILO (International Labour Organization), UNEP (United Nations Environment Programme), UNITAR (United Nations Institute for Training and Research) and WHO. The deliberations of the WPMN also includes participants from non-member countries like Russia, Brazil, Singapore, Thailand and India, and intergovernmental organisations like the WHO and the UNEP, ISO Technical Committee 229 on nanotechnologies, as well as other environmental NGOs and the Trade Union Advisory Committee to OECD. The multiplicity of actors not only underlines the importance of the topic but also illustrates the drive within the OECD to widen representation and gain legitimacy, as well as to secure support for its policy recommendations in this area.

Apart from the inclusion of several international agencies within the WPMN's deliberations, the OECD is active in the deliberations of other international organisations related to nanomaterials. It has collaborated with the IEC, NIST (US National Institute of Standards and Testing) and the ISO in holding a Joint International Workshop on measurement and characterisation for nanotechnologies.[10] One of the outcomes of the workshop was the establishment of the Nanotechnology Liaison Coordination Group to ensure coordination of activities between these international and national actors. Also the ISO has urged WPMN members to ensure that they coordinate with their national representatives on

ISO/TC (Technical Committee) 229 at the national level. Coordination has also taken place with the IFCS secretariat, when they participated in Forum VI of the IFCS, held last year in Dakar.

International Risk Governance Council

The IRGC has been working on developing an overall framework for the risk governance of nanotechnology since 2005 onwards. In 2007 it published a report (IRGC, 2007), along with survey reports targeting the role of specific stakeholders, viz. government, industry, research organisations and NGOs. Currently, the second project focuses on developing specific risk governance frameworks and techniques for nanotechnology applications in food and cosmetics. The methodology followed in both projects included the setting up of an expert body that would author the reports, in addition to inputs from multi-stakeholder expert workshops. Final recommendations from the second project were published in 2009 (IRGC, 2009).

The policy brief 'Nanotechnology Risk Governance: Recommendations for a global, coordinated approach to the governance of potential risks' identified four generations of nanotechnology products and sought to differentiate between risk governance frameworks (Frame I and Frame II) based on whether these applications contained active or passive nanostructures. The first generation products were constituted by passive nanostructures (Frame I applicable), whereas the second, third and fourth generations included active nanostructures (therefore Frame II would be applicable). The IRGC states that although risk assessment and risk management are a prerequisite for frames, current regulatory structures and processes are adequate in responding to the risk assessment requirements under Frame I. In the case of frame II, however, applications are likely to fall outside the remit of existing regulatory bodies and risk assessment methodologies may simply not exist. The report seeks to underline the differences in national regulations and the lack of harmonisation in risk assessment disciplines globally in this area as one of the shortcomings in the current scenario. The dearth of risk assessment data is identified as one of the major impediments in developing a suitable regulatory approach. It strongly supports collaborative measures between industry, governments and other stakeholders, viz. international voluntary agreements, and sees it as the future basis for undertaking regulatory action. Another important contribution of the report has been to identify and recommend action points for stakeholder groups like government, industry, international organisations, academia and civil society. It also states that the 'policy brief is targeted at policymakers engaged in planning, oversight and funding of nanotechnology regulation, research and practical applications'.

9.5   Analysis

As mentioned earlier, out of the three actors it is the OECD which prima facie seems to have the strongest linkages with regulatory agencies through its member countries. It is best positioned to secure the uptake of its policy making on nanotechnology by the regulatory agencies in its member countries. From the point of the regulatory agencies, the OECD appears to be a strategic forum in which international harmonisation efforts could be taken up. The European Commission Communication (to the European Parliament, the Council and the European Economic and Social Committee) on the Regulatory Aspects of Nanomaterials states that:[11]

> The development of standards and test methods requires close international collaboration to ensure that scientific data can be compared globally and that scientific methods used for regulatory purposes are harmonised. A main forum for the coordination of activities at the international level has been provided by the OECD Working Party on Manufactured Nanomaterials. Work is also carried out in the framework of the International Organisation.

Similarly the 'Nanosciences and Nanotechnologies: An action plan for Europe 2005–2009. First Implementation Report 2005–2007'[12] states that:

> A principal forum for the coordination of activities at the international level has been provided by the OECD Working Party on Manufactured Nanomaterials.

The Council's conclusions on nanosciences and nanotechnologies[13] also stated that it:

> Is convinced that Europe's chances of being and staying at the forefront in this field hinge upon its capacity for coordination; reiterates the need for a single Community focal point for coordination and the importance of the EU's speaking with one voice on the international stage, particularly in the light of the challenges presented by patent protection in China; calls therefore on the Commission

and Members States to devise mechanisms to effectively coordinate actions in this field; urges the Commission to take into account in its policy making all activities within the OECD (e.g. definitions, nomenclature, risk management) and UNESCO (ethics).

This is the clearest indication that not only is the EU aware of the value of the work undertaken by the OECD, but that it supports the OECD as the 'preeminent forum' for coordinating activities internationally.

The IFCS is a unique forum in terms of the equal representation that it provides to countries, intergovernmental organisations and NGOs. This, however, also complicates policy making, leading to disagreements between countries and NGOs, with the latter supporting more aggressive positions on the social and ethical dimensions of new technologies. This division also surfaced at the IFCS Session VI in discussions on the topic.[14] The important step for the IFCS was to introduce both the environmental safety and the economic aspects of nanotechnology into the international chemicals management agenda, through the Dakar Statement on Manufactured Nanomaterials. The recommendations were to be considered by ICMM2 for further action. However the rejection of the proposal by ICMM2 to incorporate the IFCS as an advisory body puts into question the future of IFCS and therefore its capacity to take the nanotechnology agenda it has constructed any further. In this context it is important mention that although nanotechnology did feature as a part of the final omnibus resolution on emerging issues,[15] there was no reference to the Dakar Statement. The final resolution on emerging issues in the ICCM2, while recognising the potential human health and environmental risks that are associated with nanotechnologies and manufactured nanomaterials, only underlined the need for coordinating information gathering between government and private sector and making it accessible to all stakeholders. Further the relevance of these issues, particularly to developing countries and countries with economies in transition were also underlined. It is therefore to be seen whether further work is undertaken through the Open-Ended Working Group that has going to be set up to carry out intersessional until the start of ICCM3.

In the case of the IRGC, there are several cleavages through which it could leverage its role as a pre-eminent actor in setting up risk disciplines in the field of nanotechnology at the global level. The first such cleavage appears to be in the selection of its expert body. In the first project the experts included inter alia; Dr Mihail Roco of the National Science Foundation and Professor Ortwin Renn of the University of

Stuttgart. Both are widely recognised subject experts in the field of nanotechnology and risk regulation respectively. Second, the risk governance frames that have been developed by IRGC have also been widely circulated by the authors in high impact forums.[16] Third, the project itself was funded by the EPA and the Department of State of the USA. The potential value of the IRCG's policy outputs increases tremendously given the probability of their adoption by the EPA. Similarly the multistakeholder expert workshops and stakeholder surveys have included wide representation from countries (policy makers), intergovernmental organisations, academicians and civil society organisations active on this issue. The IRGC is therefore a clear example of an actor adopting a strategy of legitimation by association.

A comparative look at these three actors seem to suggest the following: first, the OECD seems to be best placed to emerge as the foremost actor to take a lead on policy making on nanotechnology issues globally. The intergovernmental nature of the OECD gives it close linkages with countries and their regulatory agencies. Further the internal organisation of the WPMN in terms of steering groups with substantial responsibility to representatives of national regulating agencies has also helped in the easy adoption of frameworks and strategies that can fast-track the process of policy development. This has also made possible a substantial deepening of the global agenda on nanotechnology from that of making general statements to designing operational standards, terminology and regulatory options. Perhaps most importantly it is the support of EU member states and the European Commission that has been instrumental in the emergence of the OECD as an acceptable and therefore legitimate forum for global policy making on nanotechnology.

Second, in the case of the IFCS, one could imagine that there is substantial support for its work among NGOs and a number of developing countries – primarily because it is one of the most participatory forums at the international level. However, the proceedings at the ICCM2 suggest that the IFCS does not enjoy substantial support from a number of developed countries, which essentially sees it as replicating the SAICM. Given that the ICCM2 rejected the IFCS's proposal, it does seem difficult for it to continue functioning given the current resource constraints. Further the absence of any reference to the Dakar Statement on Manufactured Nanomaterials in the omnibus resolution of the ICCM2 seem to suggest that there is considerable opposition by developed countries to lending support to the IFCS policy making on nanotechnology. Overall it seems that there is a low probability of the IFCS emerging as a major actor in nanotechnology policy making.

Third, the IRGC has the potential to emerge as an important actor on global policy making on nanotechnology risk governance. It has been able to build and operate a network of growing stakeholders and interested parties through its periodic workshops. Also it has been able to carve a niche for itself by focusing on risk governance in nanotechnology. The choice of its second project – nanotechnology risk governance in food and cosmetics – is evidence of the focus on high priority policy making with the aim of targeting policy makers across countries.

9.6 Conclusion

The preceding discussion highlights the growing importance of actors like the IRGC, the IFCS and the OECD as acceptable forums of global policy making. It reflects several trends. First is the increasing emphasis on international harmonisation of standards and regulatory policies in specific fields like nanotechnology. Given that nanotechnology will have a potential global impact, it has been emphasised that there is a need to coordinate action globally and in a strategic manner. Second, countries are also aware of their own resource limitations and therefore are more willing to cooperate through international forums on standard setting and regulatory policy harmonisation. Third, global policy making is also a highly contested field, in which countries vie with each other in influencing policy making based on their own national standards and practices. Within the EU context, it is useful to highlight that internally private and soft regulation has been increasingly adopted across policy areas in an effort to provide guidance and enable harmonisation within the single market. The EU's Open Method of Coordination in the field of social policy highlights the network of EU institutions and private players that undertake policy making (Zeitlin, Pochet and Magnussen, 2005; Coen and Thatcher 2008). The trend towards forming regulatory partnerships with actors like the OECD could reflect the dynamics of the internal market.

It is important to stress that all these actors have displayed agency in determining their own mandate. They have been able to secure resources and partnerships with influential stakeholders in gaining acceptability for their policy recommendations. This is in sync with empirical studies that indicate the influence of transnational institutions on national policy making (Carney and Farashahi, 2006). Nevertheless, abilities on the part of these actors to negotiate with states and other international actors on their policy making tilts and priorities are imperative in determining the impact they will have on formulating normative standards which will be adopted nationally.

It is in this context that all these three actors have undertaken specific steps towards establishing policy supremacy in specific sub-fields of nanotechnology. This field is rapidly growing and given the nascent stage, any actor could gain overran overall advantage. However, as is evident from the above analysis it is crucial that there is certain convergence of interest between countries and the ambitions of these actors. Since the international regulatory field continues to be dominated by countries, it is important that a number of countries identify with the goals, roles and regulatory objectives of the actors. The EU has clearly identified the OECD to be a primary actor through which it could proceed with its harmonisation efforts in the field of nanotechnology. In the case of IFCS, however, despite it being the first actor to identify nanotechnology as an important part of the international chemicals safety agenda – it has been unable to capitalise on it. The primary reason – not unexpectedly – was the lack of support from developed countries, which regarded the forum as too open and too prone to influence from NGOs for it to gain substantial control over policy making. This seems to be the primary rationale behind their lack of support for the proposal to bring the IFCS into the fold of ICMM2 as an advisory body. First, the decision to open negotiations on nanotechnology through the OEWG and not the IFCS and second, the absence of reference to the Dakar Statement, is also evident of their intention to delegitimise the IFCS as a forum for future negotiations on nanotechnology. This also underlines the critical background role played by states in supporting certain institutional types – in this case an intergovernmental OECD over others – more open and collaborative IFCS.

Within the EU there seems to be a consensus across regulatory agencies that the OECD provides an important forum for developing standards and regulatory policy on nanotechnology. In that sense one is witnessing a gradual shift towards international co-regulation between EU regulatory agencies and actors such as the OECD and the IRGC. This partnership makes political sense, not only in terms of efficiency, but also because international organisations need to legitimise their existence by extending or deepening their mandate to cover areas and issues that are of high priority in most countries. In that sense, nanotechnology represents a priority issue, and therefore one would expect to see a greater role for such actors within this field. This trend also underlines the urgent need to investigate policy formulation processes internationally in order to delineate the field in terms of providing more transparency and legitimacy to such processes. Also it raises the question of whether the choice of single actor schemes over more collaborative ones – at the agenda-setting level – will adversely affect downstream (negotiation, implementation, monitoring and enforcement) regulatory activities in a manner that will impede their ability to fulfil common interest objectives.

Notes

1. See for instance the DEFRA (Department for Environment, Food and Rural Affairs) study, 'A regulatory gaps study for the products and applications of nanotechnologies – CB01075', April 2006. Also see European Commission, 'Nanosciences and Nanotechnologies: An action plan for Europe 2005–2009: First Implementation Report 2005–2007', 8, 2007.
2. See, Project on Emerging Nanotechnologies, Nanotechnology Consumer Products Inventory. Available at http://www.nanotechproject.org/inventories/consumer/. Accessed on 5 May 2009.
3. By this we differentiate between a mandate that is from an external source – for example, UNFCC calling upon the ICAO to consider GHG emissions from climate change as a topic for further policy making – and that which is internally driven – e.g. members of IFCS voting for inclusion of nanotechnology and manufactured nanomaterials as an agenda item.
4. Available at http://www.who.int/ifcs/en/. Accessed on 12 March 2009.
5. Available at http://business.timesonline.co.uk/tol/business/economics/article1800929.ece. Accessed on 7 March 2009.
6. Available at http://www.irgc.org/IMG/pdf/IRGC_SumInfo_24_Sept_08.pdf. Accessed on 10 March 2009.
7. For instance the OECD has recently published a new booklet on the first five years of OECD work on nanosafety. Available at http://www.oecd.org/dataoecd/6/25/47104296.pdf. Accessed on 12 March 2011.
8. IFCS/FORUM-VI/07W.
9. SAICM/ICMM.2/CRP.33.
10. Available at http://www.standardsinfo.net/info/livelink/fetch/2000/148478/7746082/index.html. Accessed on 15 April 2009.
11. COM (2008)366, SEC(2008) 2036.
12. COM (2007) 505 final.
13. 2832nd COMPETITIVENESS (Internal market, Industry and Research) Council meeting, Brussels, 22 and 23 November 2007.
14. See Summary of the Sixth Session of the Intergovernmental forum on Chemical Safety: 15–19 September 2008, Earth Negotiations Bulletin, IISD, September 2008.
15. SAICM/ICMM.2/CRP.25/Add.4.
16. See for instance, Ortwin Renn, presentation on 'Nanotechnology: Risk Governance', Second Annual Nanotechnology Dialogue, 2 October 2008, Brussels.

References

Abbott, Kenneth W., and Duncan Snidal, 'Why States Act through Formal International Organizations', *Journal of Conflict Resolution*, Vol. 42 (1) (1998), pp. 3–32.

Abbott, K. and D. Snidal, 'The governance triangle: regulatory standards institutions and the shadow of the state', in W. Mattli and N. Woods (eds), *The Politics of Global Regulation* (New Jersey and Oxford: Princeton University Press, 2010).

Barling, D., 'Regulatory conflict and marketing of agricultural biotechnology in the European Community', in J. Stanyer and G. Stoker (eds), *Contemporary Political Studies* (Nottingham: Political Studies Association of the UK, 1997), pp. 1040–8.

Berstein, S. and E. Hannah, 'Non-state global standard setting and the WTO: legitimacy and the need for regulatory space', *Journal of International Economic Law*, vol. 3 (2008), pp. 1–34.

Buonanno, L., S. Zablotney and P. Keefer, 'Politics versus science in the making of a new regulatory regime for food in Europe', *European Integration online Papers*, vol. 5 (2001) no. 12. Available at http://eiop.or.at/eiop/texte/2001-012a.htm. Accessed on 12 March 2009.

Bowman, D. and G. V. Calster, 'Reflecting on REACH: global implications of the European Union's chemical regulation', *Nanotechnology Law & Business*, vol. 4 (2007), pp. 375–84.

Bowman, D. and G. Hodge, 'Governing nanotechnology without government', *Science and Public Policy*, vol. 35 (2008) no. 7, pp. 475–87.

Bowman, D. and G. V. Calster, 'Regulatory design for new technologies: Spaghetti junction or Bauhaus?', *Notizie di Politeia*. vol. XXV (2009) no. 94, pp. 75–93.

Carney, M. and M. Farashahi, 'Transnational institutions in developing countries: the case of the Iranian Civil Aviation', *Organization Studies*, vol. 27 (2006), pp. 53–77.

Cantley, M., 'The regulation of modern biotechnology: an historical and European perspective', in D. Brauer (ed.), *Biotechnology: Legal, Economic and Ethical Dimensions* (Weinheim: VCH, 1995), ch. 18.

Chambers, G., 'The BSE crisis and the European Parliament', in C. Joerges and E. Vos, *EU Committees: Social Regulation, Law and Politics* (Oxford: Hart Publishing, 1999), pp. 95–106.

Coen, M. and M. Thatcher, 'Network governance and multi-level delegation: European networks of regulatory agencies', *Journal of Public Policy*, vol. 28 (2008), pp. 49–71.

Dorbeck-Jung, B. D. Bowman and G. V. Calster, 'Governing nanomedicine: lessons from within, and for the EU medical technology regulatory framework', *Law & Policy*, vol. 33 (2011) no. 2, pp. 256–75.

Environment Protection Agency, *Nanoscale Materials Stewardship Program: Interim Report*, 2009. Available at http://www.epa.gov/oppt/nano/nmsp-interim-report-final.pdf.

Felcher, E. M., 'The Consumer Product Safety Commission and Nanotechnology', *Project on Emerging Nanotechnologies* (Washington: Woodrow Wilson International Centre for Scholars, 2008).

Fielder, F. and G. Reynolds, 'Legal problems of Nanotechnology: An overview', *Southern California Interdisciplinary Law Journal*, vol. 3 (1994), p. 593.

Fraiberg, J.D. and M.J. Trebilcock, 'Risk regulation: technocratic and democratic tools for regulatory reform', *Mc Gill Law Journal*, vol. 43 (1998), p. 846.

Francioni, F., 'Genetic resources, biotechnology and human rights: The international legal framework', *European University Institute Law Working Paper*, vol. 17 (2006), pp. 3–8.

Gartner, S., J. Kullmer and U. Schlottmann, 'Chemical safety in a vulnerable world', *Angewandte Chemie International Edition*, vol. 42 (2003), pp. 4456–69.

Hackett, E. J., O. Amsterdamska, M. Lynch and J. Wajcman (eds), *The Handbook of Science and Technology Studies*, 3rd edn (Cambridge MA: MIT Press, 2008), pp. 63–7.

IRGC, 'Nanotechnology risk governance: recommendations for a global, coordinated approach to the governance of potential risks', *Policy Brief* (Geneva, 2007).

IRGC, "Appropriate Risk Governance Strategies for Nanotechnology applications in food and cosmetics", *Policy Brief* (Geneva, 2009).

Jasanoff, S., 'Product, process, or programme: Three cultures and the regulation of biotechnology', in M. Bauer (ed.), *Resistance to New Technology* (Cambridge: Cambridge University Press, 1995), pp. 311–31.

Jones, K., 'The WTO core agreement, non-trade issues and institutional integrity', *World Trade Review*, vol. 1 (2002), p. 257.

Lin, A., 'Size matters: regulating nanotechnology', *Harvard Environmental Law Review* vol. 31 (2007), p. 349.

Mattli W. and N. Woods, 'In whose benefit? explaining regulatory change in global politics', in W. Mattli and N. Woods (eds), *The Politics of Global Regulation* (New Jersey and Oxford: Princeton University Press, 2010).

Marchant, G. E. and D. Sylvester, 'Transnational models for regulation of nanotechnology', *Journal of Law, Medicine and Ethics* (2006), pp. 714–25.

Mehta, M., *Regulating Biotechnology and Nanotechnology in Canada: A Post-Normal Science Approach for Inclusion of the Fourth Helix*, 2(2), 2002. Available at http://www.nanoandsociety.com/ourlibrary/documents/mehta-nus-paper2002.pdf. Accessed on 24 February 2009.

Wilson, R. F., 'Nanotechnology: The Challenge of Regulating Known Unknowns', *The Journal of Law, Medicine & Ethics*, vol. 34 (2006), pp. 704–13.

Wilson, R. F., 'Nanotechnology: the challenges of regulating known unknowns', POSTERS, *Nanomedicine: Nanotechnology, Biology, and Medicine*, vol. 2 (2006), pp. 313–8.

Zeitlin, J., C. Pochet and L. Magnussen (eds), *The Open Method of Coordination in Action: The European Employment and Social Inclusion strategies* (Brussels: Peter Lang, 2005).

10

The European Patent System: Emerging Technologies, the Patent Quality Challenge and Mechanisms to Deal with it

Evisa Kica

10.1 Introduction

This chapter provides a comprehensive analysis of the European patent system and the current challenges surrounding the quality of patents. Patents are an essential element of economic growth that foster innovation through the generation of new scientific and technological knowledge. The impact of patents to spur innovation and economic growth in Europe has been clearly stated by the European Commission and Council (e.g. European Commission, 2008; Council of the European Union, 2009). The Lisbon Agenda of the European Council also stresses the role of IP protection for advancing European competitiveness and growth. Patents continue to be used increasingly by private companies and public research organisations to protect new inventions, which have created new waves of innovation and prompted shifts within the legal and regulatory framework of patent regimes (OECD, 2004). However, the emergence of new technologies and the increase of patent applications have all impacted the IP regulatory framework and the ability of patent examiners to issue high-quality patents.

Researchers claim that in recent years the innovation trends and functioning of the patent system have been challenged by different factors, both exogenous and endogenous (Cowan et al., 2007; Guellec and van Pottelsberghe de la Potterie, 2007). The emergence of new technologies and scientific advancements (e.g. biotechnology, computer-related inventions, nanotechnology) are perceived as the main exogenous forces. These technologies have posed many questions to the scope of patentable subject matter and generated the need for IP legislators to consider regulating new types of technology and knowledge. For instance, several patent offices (including the European Patent Office) have taken a leading role in expanding the scope of subject matter for which patent protection may be obtained, adding biotechnology as a patent eligible subject matter within IP institutions (Lee, 2005). As a result, the number of patent applications for emerging technologies and the inclusion of new actors within the system (i.e. high-tech firms, research institutions, start-ups and networks) have grown exponentially, challenging the future functioning of the European patent system (van Pottelsberghe de la Potterie, 2009). Among the main concerns raised by policy makers, the business community, practitioners, academics and examiners, is the extent to which patent examiners have adequate resources and expertise to perform their search and examination tasks, and issue high-quality patents.[1]

The quality of granted patents is considered the main endogenous factor that is challenging the ability of the patent system to encourage innovation and the diffusion of technology (Cowan et al., 2007; Hall, 2007). There are various definitions of patent quality. While scholars look at the issue from a broader perspective, relating it in principle to the extent to which patents fulfil the patentability criteria (novelty, non-obviousness and industrial applicability), users of the system claim that patent quality is related to the costs of patenting, timeliness of examination and the performance of the products provided to the customers. However, patent authorities link the quality of patents with the successful disclosure of the patent application by the applicants (Scellato et al., 2011). In this chapter, patent quality issues refer to the search and examination processes, as well as the ability of patent applications to fulfil the patentability requirements.

Concerns about the quality of patents have generated many debates. The first comprehensive studies in this direction started in the United States (e.g. Jaffe and Lerner, 2004; Bessen and Meurer, 2008). However, the European patent system cannot be excluded from the risks and negative implications of low-quality patents (van Pottelsberghe de la Potterie, 2009). The complex relationship of patents with other important areas in Europe, such as European competition policy, technology standards policy, international trade and health care policy, has made IP rights policy crucial at the European level (Scellato et al., 2011). Low-quality patents do not comply with the patentability criteria and may seriously harm the innovation process by causing the public to pay supracompetitive prices for products (due to high royalty costs for licences) (Lemley and Shapiro, 2005). They can also impose unnecessary constraints on downstream innovation that leads to fewer incentives for technology developments and generate 'patent thickets', where overlapping patents are created and prevent market entry by new innovators. Most importantly, patented inventions of low-quality, which do not comply with the patentability standards, risk undermining the scientific research and the advancement of technologies that could benefit both the innovation process and the public needs (e.g. medical advancement, patient care etc). As such, patent quality issues have been addressed by many scholars (Burke and Reitzig, 2007; Edfjäll, 2007; Elsmore, 2009; Shang, 2009; Wagner, 2009; White, 2004), policy makers (Cowan et al., 2007; European Commission, 2008) and patent experts (Harhoff, 2009; Hammer, 2010). However, there has been less analysis of the patent quality mechanisms used in patent offices to tackle the issue of low-quality patents and foster technology developments.

The aim of this chapter is to present the issue of low-quality patents in the European patent system and to provide comprehensive evidence of the administrative mechanisms that are put forward to tackle this issue. In particular, the chapter aims to provide an answer to the following questions.

First, what are the major challenges that the current system is facing in responding to the developments of new technologies? Second, what mechanisms do patent authorities and scholars put forward to enhance the quality of patents and overcome the large number of patent applications and backlogs? Third, how we can apply what we have learned from the current mechanisms and challenges to enhance further the ability of the patent system to promote high-quality patents?

The remainder of this chapter is organised as follows. Section 10.2 addresses the regulatory framework of the European patent system and the major challenges that this faces. It discusses both recent patentability developments and trends. Section 10.3 examines the scholarly debate of the mechanisms that support patent quality enhancement through improvements in the examination process. Section 10.4 displays the landscape of the administrative mechanisms at the European Patent Office (EPO). Finally, section 10.5 presents conclusions and policy recommendations aimed at fostering the better functioning of the European patent system and the value of patents.

10.2 Reconsidering the protection of intellectual property at the European level

The initiatives towards the establishment of an IP protection regime at the European level were finalised in the Munich Convention of 1973. During the 1970s, the European Patent Convention (EPC) established the EPO as an alternative through which countries could acquire

IP protection. The European patent system is not exclusive; applicants can also obtain strictly national patents if they are interested in obtaining protection only in one or a few of the EPC contracting states. However, European patents are considered to be the most effective way for inventors to protect their inventions outside their national boundaries through a centralised patent-granting process at the EPO (Graham et al., 2001). In line with Article 52 (1) EPC, the EPO grants patents to any invention, in all fields of technology, provided that they are new, involve an inventive step that is not obvious to a person skilled in the art, and are industrially applicable (Tödtling and Trippl, 2005). After patents are granted by the EPO, the legislative authority for the use of patents remains under the national jurisdictions of the contracting states (Friebel et al., 2006). European patent applications are examined by the EPO's Examination Division, which is staffed by technical experts. The EPO's Search Division draws up a Search Report relevant to the subject matter of the patent claim[2] (Akers, 2000). In addition, a great number of patent applications undergo substantive examination by the Examination Division.[3] In addition, EPC provisions recognise the contribution of external actors in the patent prosecution process as well. The EPC has laid down Article 115, which provides that after the publication of the European patent application, third parties can communicate certain information or documents to the examiner in charge concerning the patentability of the invention for which an application has been filed. No fees are required for the submission of observations, and the person filing an observation may not be a party to the proceedings before the EPO (Guellec and van Pottelsberghe de la Potterie, 2007). However, this mechanism continues to be used infrequently. The scope of the patents is mostly determined by highly specialised, technical in-house examination following the regulatory process, which is disguised behind the technical patentability standards. After a patent is granted, parties may file their opposition to the patent pursuant to Article 99 EPC or an appeal with the Technical Boards of Appeal and, in certain cases, may file a petition for review with the Enlarged Board of Appeal.

Confined as it is to this world of highly specialised, technical in-house examination, the EPO enjoyed a relatively quiet life until the 1990s (Borrás, Koutalakis and Wendler 2007). After this period, certain patentability controversies started to challenge the functioning of the EPO and the ability of patent examiners to assess new technologies. From 1998 to 2007 the number of patents filed within the EPO grew from 82,300 to 140,700 (WIPO, 2008). Concomitant to this was the increase in the average size of patent applications, which have more than doubled, including higher number of claims with broader content (in 1980s a patent typically included 10 claims, while the average is now 22). Niel Stevnsborg and Bruno van Pottelsberghe de la Potterie (2007), add to this debate claiming that in recent years they have observed a tendency for patent applicants to use 'new filing strategies'. First, patent applicants file 'doubtful' applications with broad claims and merge several patents into one application, hiding in this way the real invention among a myriad of claims and pages.[4] Second, applicants file divisional applications[5] and split the patent into smaller subsequent patents, which are of low quality. Third, it is interesting to note that when dealing with new technologies, applicants are often reluctant to communicate all relevant prior art information to the EPO examiners. They usually wait for the examiner's Search Report and then consult the prior art themselves. In fact, Rule 42(1)(b) of the EPC states that patent applicants should indicate the background art in their patent application, which as far as known to the applicant, can be regarded as useful information for the examiners to understand the inventions, and draw up the European Search Report and examination. However, the EPC provisions are not interpreted as placing an obligation on applicants to communicate to the EPO examiners every item of prior art information relevant to the application. This has placed further stress on examiners to assess the patentability of the claimed inventions, since there is either too little information about inventions' prior art (e.g. computer software patents) or too much (e.g. biotechnology patents), and examiners lack the means to sift through it (Noveck, 2006).

Another worrying fact surrounding the European patent system is the quality of patents granted by the EPO. Within the current system, examiners are highly encouraged to work faster to reduce the patent pendency rates and increase productivity. However, the severely limited employee schedule (10–15 hours per application[6]) means that examiners often fail to proceed quickly with patent applications (which leads to huge backlogs) or to consider all relevant prior art information. Most importantly, prior art on new, emerging technologies cannot be found easily and the evaluation of the search results for these technologies requires an enormous number of senior-level staff, which obviously is far beyond the working capacity and timeframe within which examiners operate (Ganguli, 2001). As a result, the EPO is being challenged by an increasing number of oppositions against the validity of granted patents (Holzer, 2005).

Indeed, the patent system was not designed to operate in such an environment of booming applications, huge backlogs and oppositions, but the system continues to hold to its objective of promoting innovation by granting exclusive rights only to inventions that contribute to the promotion of technological innovations and the diffusion of knowledge (van Pottelsberghe de la Potterie, 2009). To this aim, patent scholars have proposed different mechanisms that can be used by patent offices, examiners and applicants to foster the quality of patents and the functioning of the European patent system.

10.3 Enhancing the quality of patents: Review of academic debate

Scholarly debate on patent quality reveals that there is no one single method for ensuring higher patent quality, but it does distinguish five broadly defined mechanisms that provide for an improved patentability environment: an incentive-based approach to improve the quality of patents, administrative changes, patent law changes, better patent information to patent applicants and examiners, and technical advancement for patent examiners.

An incentive-based approach to improve the quality of patents

Perhaps the most complete conception of the incentive-based approach is set forth by Polk Wagner (2009), who supports the argument that patent offices and examiners must draft appropriate incentives to encourage high-quality patents. To this aim, patent authorities should first create a strong 'multidisciplinary analytical capability' to assess management needs and technology developments, and guide patent examiners to emphasise patent quality rather than quantity (Merrill, Levin and Myers, 2004). In the same vein, Matthew Elsmore (2009) claims that multidisciplinary analysis and incentive mechanisms are significant for helping patent law to cope with new technology and innovation requirements. Furthermore, patent scholars argue that an incentive-based mechanism will encourage patent offices and officials to keep experienced examiners (van Pottelsberghe de la Potterie, 2009), to hire external expertise on technologies that pose problems (White, 2004) and to penalise patent holders, i.e. with cash fines and 'infectious invalidity' in time or costs (Elsmore, 2009).

Administrative changes

Following the incentive-based approach to improving patent quality, Wagner (2009) places administrative reforms at the centre of the patent-quality mechanisms that can be used by patent offices to reduce incentives to defer the quality of patent claims during the prosecution process. In particular, he argues that to achieve effective administrative reforms, patent offices should introduce supportive or financial means to increase the number of patent examiners in specific technical fields and to encourage feedback on low-quality patents. Katherine White (2004) adds to this debate by asserting that supportive means are also important for helping applicants to make their applications public and to provide for 'concise and precise' claims when filing patent applications. However, the analyses of Stephen Merrill, Richard Levin and Mark Myers (2004), Roger Shang (2009) and Bruno van Pottelsberghe de la Potterie (2009) move beyond these arguments. These scholars identify other strategies for improving patent quality, central among these being the examination guidelines and processes. They argue for an improved examination and pre-granting opposition process and for better quality assurance techniques. In this respect, Shang (2009) contends that the inclusion of third parties (i.e. *inter partes* re-examination and post-grant reviews) in the examination process provides added value to the assessment of the validity of questionable patents.

Patent law and organisational changes

An increase in the number of patent applications, financial means, quality assurance techniques and validity claims cannot be achieved without substantive changes in patent law. To respond to the challenges of new technologies and enhance the quality of patents, patent offices should provide for efficient legislative actions and reconsider their regulatory frameworks, patentability standards and organisational structures (Cowan et al., 2007). Regarding new technologies, Elsmore (2009) and Merrill, Levin and Myers (2004) suggest the improvement of the requirements for defining the patentable subject matter, and the international harmonisation of patent examination procedures and standards. Under these mechanisms, patent examiners and offices would be able to achieve mutual agreements about the patentability of new technologies. However, because patent law is a specialised field with many active players, high-quality patents will result only if the patent offices balance the interests of active and passive users, and legislate an 'open review procedure' that allows third parties to challenge patents (Guellec and van Pottelsberghe de la Potterie, 2007). Additionally, patent offices should improve their management structures and accountability, and reformulate their patent information policies (Edfjäll, 2007).

Technical advancement of examiners

Recent work by Bruno van Pottelsberghe de la Potterie (2009) indicates that backlogs and the falling quality of patent applications can be easily reduced through training schemes that foster the performance of patent examiners. In this respect, patent offices need to establish qualification mechanisms (i.e. tests, ongoing examinations and coaching services) and recruit or promote examiners based on their relevant skills.

Better patent information to patent applicants and examiners

Academic literature indicates that the assessment of patents is most often associated with the clarity of information in patent claims and the examination procedure. Patent information on claims, provided by applicants, contributes to the clarity of patents and leads to a more cost-effective examination process (Cowan et al., 2007; Harhoff, 2009). Therefore, patent offices should allocate additional resources to examiners and ensure ongoing deliberations between applicants and patent authorities on the patentability of various subject matters. Other scholars claim that patent examiners' access to literature (i.e. scientific and patent literature) and collaboration with commercial patent information providers or other institutions specialised in protecting certain industries also provides for high-quality outcomes (White, 2004; Elsmore, 2009). Paul Burke and Markus Reitzig (2007) and Curt Edfjäll (2007) contribute to this debate and state that applicants will be able to conduct thorough claim-construction analyses only if patent offices disseminate all data collected to the public, publish all patents in force and encourage better cooperation among information providers and information users.

Taken together, these mechanisms and arguments are of crucial importance as they provide us with more opportunities to understand how modest reforms can be made to encourage high technological quality and sustainable property rights (Burke and Reitzig, 2007). Following the scholarly debate on patent quality, a variety of mechanisms have also been endorsed by the Administrative Council of the EPO. The patent quality mechanisms of the EPO develop within similar lines as the academic debate, but in addition to administrative and procedural changes, it emphasises collaboration between patent offices, the development of quality standards and the utilisation of prior-art information from other sources.

10.4 The landscape of the EPO mechanisms to enhance the quality of patents

In 2007 the Administrative Council of the EPO and its Board (i.e. Board 28[7]) started to work on the issue of patent application backlogs and examiners' workload. On the basis of its findings, Board 28 proposed five strategic recommendations to improve the quality of patents and reduce backlogs at the European level. These strategies can be divided into two groups. In the first group are the strategies that address the quality of search and examination procedures, and the EPO's utilisation of work from other patent offices and sources. In the second are strategies that deal with improving the quality of products and processes at the EPO.

Raising the bar at the EPO

The 'Raising the Bar' initiative was implemented in 2007 with the initial objective to:

1. support the EPO's practice on the assessment of the patentability requirements;
2. ensure that examiners effectively conduct search and examination only on the subject matter for which a protection is sought;
3. formalise the current practices within the system for applicants to identify the basis of patent claim amendments; and
4. restrict time limits for the filing of divisional applications.

In this way, the 'Raising the Bar' initiative aimed at responding to the devaluations of European patents and the inadequacies of the European patent system (Hammer, 2010). To achieve these objectives, Board 28 provided three recommendations:

(A) Changes to the existing practices and procedures: In this respect, the Board proposed that the EPO make crucial changes to its legislative framework, focusing on the practice and procedures of granting patents. To enhance the quality of patents, the Board argued that the EPO should primarily reconsider the requirements for granting patents and assessing the inventive step. At this point, the following measures were recommended: amendments of the search and examination guidelines, assessment of the requirements for the 'technical character' of an invention requirement, expansion of the scope of actors involved in decisions to grant/refuse patent applications, and consideration of the possibility to make patent applications or parts of them open to public inspection (European Patent Office, 2007).

(B) Expanding the scope of actors involvement in the patent prosecution process: Since the patent system is a complex functioning system that is based on heterogeneous actors (i.e. applicants, their representatives, third parties), effective and legitimate outcomes will be ensured only

if these actors are included within the patent prosecution process. According to Board 28, the EPO could ensure a more inclusive environment within the patent prosecution process if it:

1. stimulates applicants to search the inventions and to explain the basis of patent claim amendments;
2. establishes mandatory requirements for the applicants to respond to the Extended European Search Report (EESR), European Search Opinion (ESOP) or Written Opinion of the International Searching Authority (WOISA) when entering the substantive examination phase;
3. introduces the top limits of claims per application;
4. collaborates with the Institute of Professional Representatives before the EPO (*epi*) to set the Code of Practice; and
5. allows third parties to request accelerated examination ('PACE' programme[8]) (Alge, 2008: 20).

(C) Changes to the EPC rules: Recommendations to change the legal standard of the EPC address mostly the level of the inventive step, which is considered to be low at the EPO. Several strategies have been proposed to achieve a higher standard of inventive step through legislative changes. First, the EPO should carefully assess (through a broad consultation process carried out at the EPO) if there is a compelling need for a legislative change by reviewing the EPC's existing proposals. Second, it should determine how the definition of the 'person skilled in the art' could be modified. Finally, the Board recommends that the EPO should assess if specific bases should be established to amend patent applications prior to the search and examination process (European Patent Office, 2010).

Following the long discussions within the EPO Administrative Council, the first package of the 'Raising the Bar' measures entered into force in April 2010. These measures introduced new guidelines to help examiners assess the inventive step, apply the EPO standards in an efficient and consistent manner, and foster communication with the applicants prior to the substantive examination process (i.e. Guidelines for Examination and the Internal Instructions) (European Patent Office, 2010; Hammer, 2010). Furthermore, the conclusions of the 'Raising the Bar' initiative reflect changes to the EPC rules. The new amended EPC rules obligate patent applicants to respond to the search opinion of the EPO before the substantive examination process (Rule 70a and 161 EPC), file divisional applications only during drastically shortened periods (Rule 36 (1) EPC), communicate with examiners during the pre-search stage (Rule 71 (3) EPC) and clearly describe the basis for amendments (Rule 137 (4) EPC). In addition, the amended rules have put further restrictions on the subject matter that will be searched by the EPO (Rule 62a, 63 and 64 (1) EPC) (Administrative Council of the EPO, 2009).

An enhanced collaboration among patent offices-building the European Patent Network

Collaboration and enhanced partnerships among patent offices are perceived by the Council of the EU as the right mechanisms to improve the quality of patents add value to the work of the EPO and strengthen the capacity of national patent offices (NPOs) to enhance the quality of the patent system in the future (Council of the European Union, 2009). As mentioned earlier, the European patent system is not the only route through which inventors can obtain exclusive rights for their inventions. Patent applicants can choose the national route if they are interested in obtaining patent protection in one or few of the EPC countries. However, when acquiring patent protection in more than three countries it is cheaper for applicants to go for a European patent rather than independent applications in several jurisdictions. Therefore, the European filing procedure has become attractive, especially for applicants that aim to sell their goods and services in international markets (Graham et al., 2001). It is important to mention that if applicants want to enter the regional European phase they have to claim the priority of the national patent application, indicating the date, the country and the file number of the earlier application (Article 88 EPC). The collaboration among NPOs and the EPO on how the search was conducted and which documents/prior art have been consulted is crucial for the second filing office (in this case, the EPO) to determine the need for further searches (Council of the European Union, 2009).

In 2005 the Administrative Council of the EPO responded to these needs and held regular discussions on improving the collaboration between patent offices in Europe. This resulted to the establishment of the European Patent Network (EPN). The basic idea of the EPN is that the patent system should be perceived as a process that builds upon the network of NPOs and the EPO, which should play complementary roles and work together to foster innovation and economic growth. The EPN builds upon four pillars: the EPO's utilisation of NPOs' search work, the establishment of a European Quality System (EQS), user support activities provided by NPOs rather than the EPO, and a new cooperation policy based on partnership. According to Jürgen Schade (2009), utilisation is considered the most important pillar for assisting patent offices to perform their roles better, avoid unnecessary duplication of work and combine efforts (knowledge and skills) in cases when cooperation is needed. Within the EPN, utilisation consists of the Utilisation Pilot Project (UPP) and the EQS.

EPO's utilisation of work from NPOs

The utilisation schemes within the EPN promote the idea of reusing the work performed at an office of first filing (i.e. an NPO at the EPO contracting state), while examiners at the office of second filing retain full responsibility for the file. To test the EPO's utilisation of the searches done by other NPOs, in 2008 the UPP was established. It analysed more than 10,000 applications coming from Austria, Denmark, Germany and the UK. The proceedings of the project were reported by the Supervisory Board of the UPP to the Administrative Council of the EPO. In the beginning of the project a target was set to receive at least 1500 applications, but a much lower number of UPP applications were submitted to the EPO. According to the evidence provided by the German Trademark and Patent Office[9] (DPMA), the UPP attracted a low level of applications, because participants did not agree with the concept of utilisation and preferred to continue with two independent search and examination processes (Schade, 2009: 628).

To develop further the utilisation of work and the collaboration between NPOs and the EPO, the management of the EPO has recommended various actions that patent authorities should consider. First, the EPO should develop better methods to identify that an application was first filed at the NPO. Second, NPOs should join forces to harmonise and standardise the search and documentation process, and develop common tools for an efficient exchange of search reports (Alge, 2008).

Furthermore, Board 28 argued that the EPO should also focus on the fee instruments when developing the concept of utilisation. Such a fee instrument could give applicants the opportunity to choose between informing the EPO of earlier searches (for which a normal fee would be applied) or request a second/independent search without providing such information (for which a cost covering fee could be charged) (European Patent Office, 2007).

EPO's utilisation of information from other sources

Utilisation of information available to other sources is crucial to the issue of patent quality since other parties (i.e. third parties that do not participate in the patent prosecution process) can provide examiners with information that is crucial for them to understand the inventions, and draw up the European Search Report and examination (Cowan et al., 2007; Edfjäll, 2007). As mentioned earlier, to supplement the ability of the EPO examiners to locate relevant information on patent applications, the EPC has laid down Article 115. In line with this, external contributors (i.e. third parties) can provide input to the search of prior art and communicate relevant information to the examiner in charge. However, within the EPO only a limited number of cases have made use of Article 115.[10]

Based on the potential impact that the input of third parties has on the quality of patents, Board 28 recommends that the EPO should improve its examination process by means of third-party reviews (European Patent Office, 2007). Such a system, in which third parties can contribute to retrieving prior art and suggesting it to the examiners, was established in 2007 at the United States Patent and Trademark Office (USPTO) in the Peer-to-Patent system. Under this system, third parties can provide additional prior art information to examiners for two months after the patent application has been published. They can post a reference or notice of prior art that they believe is relevant for assessing the patentability of specific applications. Afterwards, these references are launched on a public-view page, where others are able to view and comment on them and vote for the most relevant references for the examiners to consider (Noveck, 2006). Patent authorities in Europe should make use of the evidence from the USPTO and include the contribution of outside expertise within the patent process.

Developing a European Quality System within the EPO

The EQS is an attempt to set well-defined quality standards that should be recognised by NPOs and the EPO. The primary objective of the EQS is the constant improvement of the quality of products (i.e. patents) and services (i.e. search and examination) of the participating patent offices. To fulfil this aim, the EQS consists of two parts: the European Quality Management System (EQMS), and the Product Quality Standard (PQS). The quality management system is important for both the patent granting and examination procedure. It ensures that the organisation is aware of the customer requirements and of how products and processes meet these requirements (Philpott, 2006). Therefore, to enhance the ability of EPN members to provide high-quality services and legal certainty, the EQMS defines the common quality standards that these participatory states should establish. It sets out the basic requirements for improving: the management of resources and administrative workload, patent quality review mechanisms and quality assurance, communication between offices and other users on the search process, and requirements on the standards of the search results (EQS Working Party, 2007). The PQS, on the other hand, was established in 2008 by the Administrative Council of the EPO to ensure high-quality products. It defines the minimum standards that patent authorities should follow to classify patent applications, draft reports and communicate search results, and sets out the requirements for rejected and granted patents (Kaltenbach et al., 2008).

Over the course 2009 and 2010, the EPO has taken various efforts to implement a quality management system. It has deployed a number of internal and external quality auditing techniques to assess the quality of its search and examination procedures, and to prepare for ISO 9001 certification.[11] The EPO has conducted both a systematic and random quality control of search and examination process to determine the gaps of the system and the future actions that the office should undertake to become fully ISO-compliant.[12] In addition, corrective actions have been undertaken with respect to the quality of the search process, communication with the primary users (e.g. the Manual of Best Practice[13]) and issues of patent clarity and added subject matter (i.e. Arts. 84, 123 EPC).

10.5 Bringing theory and practice together: Challenges and the way forward

The theoretical framework and the EPO's approach to patent quality mechanisms have made it clear that the debate over patent quality is intensifying, as too are the difficulties of enhancing the quality of patents with a single approach or method. Delivering high-quality products and processes continues to be a tough task that requires the effective functioning of substantive patent law, and of procedural rules and practices, all of which are being challenged by new developments in technology. For instance, biotechnology, nanotechnology and computer-related inventions have started to become apparent in patentability trends, public and private investments in R&D and the innovation strategies of start-up companies. Since the influence of these technologies is likely to increase in the future, it is crucial for the patent system to establish better expertise and regulatory guidance to assess validity and the threats of future technology developments.

With respect to new technologies, high-quality outcomes relate closely to the quality of the examination process and the collaboration of patent stakeholders (i.e. applicants, examiners and patent offices) especially with regard to searching and retrieving all relevant prior art, the correct assessment of novelty and obviousness, and the validity of granted patents. However, in view of the current flood of patent applications, backlogs and oppositions, we suggest that more emphasis should be placed on the former aspects. Substantive search and examination is crucial to high-quality patents as it represents the 'first stage' of the technology assessment, providing the regulators and other actors with the opportunity to 'control, confine and channel *ex-ante*' the operations and successful practice of inventions (European Patent Office, 2007; Brownsword, 2008).

Similar arguments, which support high-quality patents through improvements in the examination process, can also be found in recent empirical studies of patent quality. Between 2006 and 2009 the Staff Union of the European Patent Office (SUEPO) conducted several surveys and interviews with EPO staff and patent experts to determine the main issues related to the quality of the patent system and its future improvement (Friebel et al., 2006).[14] The results from the SUEPO surveys support the need for high-quality search and examination work, based on high-quality patent information service, higher standards of patentability and appropriate time and training for examiners to master the workload and achieve high-quality results. A recent study funded by the EU Commission (Scellato et al., 2011) also reveals that there is a strong relationship between pre-grant examination activities and the likelihood for patent quality improvement. Mechanisms to maintain the skills of the examiners, to increase the participation of the third parties to aid examination and contribute to a better search of prior art, to exchange information among the NPOs and EPO examiners, and to establish higher inventive step for high-tech fields, are perceived as being the most effective patent quality mechanisms.

To sum up, it is important to note that the EPO should contribute more to improving its pre-grant patent management and be more aware of the consequences that its outputs (i.e. patents) may bring to the innovation process and to society at large. This requires patent authorities and applicants to commit to the best functioning of the patent system and put forward the use of certain measures:

(A) Change the internal culture about the mission of the office and encourage examiners not only to grant patents but also to refuse patents: Among the main challenges within the current system is to guide examiners towards adopting a friendly attitude towards applicants in order to enhance communication, and to avoid creating commercial relationships with the patentees. According to Francis Hagel (2004) patent examiners should not consider applicants as 'customers'; patents should also not be considered as products but as property rights. Certainly, the mission of the EPO is to function as a patent-granting authority, but this does not mean that the grant of patents should become a norm and rejection the exception. Neither the EPO nor examiners should depict their activities as 'business' or as providing services to applicants. To avoid these issues, patent authorities should provide incentives for patent examiners to correctly assess patent applications even if it would lead to rejections. Such incentives may include rewards to examiners per patent rejection, based on the time and effort that such refusals may require, or rewards for every substantial communication between examiners and applicants during the examination process (Stevnsborg and van Pottelsberghe de la Potterie, 2007).

(B) Establish the right mechanisms to keep the skills of patent examiners: In Europe, the quality of patents relates closely to the skills of the 3500 engineers and scientists who work as patent examiners at the EPO and exercise their discretion to grant or refuse patent applications. The examination of patents is a complex, technical and legally binding process, which requires examiners to have experience and constant training in order to deliver high-quality services (van Pottelsberghe de la Potterie, 2009). At this point, Ian Cockburn, Samuel Kortum and Scott Stern (2003) suggest that the patent authorities should follow the management literature, which promotes the idea that patent offices should create a corporate culture in the form of informal rules, common values and examples of best practices to advance examiners' ability to exercise their discretion effectively in granting patent rights With respect to new technologies, patent offices should converge their human resource practices, including training schemes, recruitment policies and individual performance indicators, and strategies for keeping the current skills of examiners. Furthermore, examiners should be encouraged to look beyond the search and examination process and analyse in more detail the work of the industry and courts.

(C) Ensure ongoing deliberations among patent offices, examiner and applicants: By definition the patentability of new, emerging technologies is still in its formation and patent stakeholders often face dilemmas about fully understanding or putting specific limitations on the functioning of these technologies. Therefore, ongoing deliberations among applicants, examiners and other stakeholders who hold various positions in relation to an application, patentee or patent (e.g. supplier, customer, licensee and investor) are crucial. Biotechnology and other new technologies affect a broader range of stakeholders. Thus, it is important for the system to promote an inclusive process of deliberation and open interchange among stakeholders to enhance the scope of information flow and understandings of the burdens and benefits of new technologies. With respect to the examination process, it is important to note that patent examiners should enhance their communication with applicants during the process and provide detailed and preliminary reports on the patentability of inventions to allow examiners to make early amendments on their applications in the matter of novelty and inventive step.

(D) Increase requirements for higher inventive step: The flood of patent applications within the last decade have brought many challenges within the system with regard to striking an appropriate balance between honest inventors who provide for new patent claims and inventors who file broad claims with less or no contribution to the state of the art. The patent system should be neither too selective (reject 'good applications') nor fairly selective (approve applications with no clear contribution to the innovation), but it is difficult for patent offices to decide where the balance should be struck. European patent law emphasises that the inventive step is a technical step which should be applied uniformly across technical fields (van Pottelsberghe de la Potterie, 2009). This, however, is difficult in practice since technologies differ and the inventive step is assessed differently based on the extent to which new inventions advance from the prior art in certain industries (Hunt, 2004). For instance, in new emerging technologies (e.g. biotechnology or information technology) with higher opportunities for research productivity, even the smallest efforts could lead to significant inventions, whereas in other mature technologies (e.g. chemistry or mechanics) with less technical and research opportunities, significant effort will only result in minor improvements. As such, the inventive step should be higher in the former case, since the technology evolves faster and low inventive steps would encourage only simple improvements to supersede other inventions, cause patent thickets and block the development of other inventions (e.g. cumulative innovations).

(E) Set a coercive method for obtaining evidence from the applicants: Since the current system does not oblige patent applicants to disclose every item of prior art information relevant to the application,[15] examiners cannot lawfully reject an application on the grounds of poor disclosure (Akers, 2000). In this way, applicants can abuse the system and file low-quality applications, which are too general or vaguely explained. These filing strategies and the resulting patents jeopardise other inventors and the quality of the examiners' work, and create huge litigation problems. The EPO could avoid this phenomenon by setting stricter procedural requirements that would oblige applicants to disclose and describe all relevant prior art information and other referred material to the examiners. These requirements would keep patentees responsible for respecting their duty of candour and acting in good faith in their dealings with the EPO.[16]

(F) Broadening users' access to the existing prior art information: As mentioned earlier, the striking growth in the number of patents over the last decade shows how innovation processes have become more competitive and dependent on new actors (e.g. new high-tech firms, research institutions, universities) that have had a remarkable impact on the development of new subject areas (such as science-based, pharmaceutical-related inventions). However, in comparison with large firms, which have established well-developed techniques to assess prior art information, these new actors lack appropriate means to search information regarding the prior art of the subject matter. The lack of resources creates fewer chances for them to file high-quality patents and defend their inventions (Tang, John and Daniel, 2001). Patent authorities should respond to these challenges and make the current databases and search tools more public and offer free online access to patent and non-patent literature.

(G) Include external actors within the examination process: The growing complexity of new inventions has made it more difficult for examiners to understand these technologies, because much of the prior art is widespread among the public or other actors specialised in the field rather than in patent databases. However, the main weakness of the current system is its inability to open the examination process effectively to external actors who could contribute to high patent outcomes. Article 115 of the EPO is not frequently used, probably because third parties (especially private companies) do not want to forewarn applicants of a competitor's interest, so they wait until a patent has been granted and then file their opposition. The EPO's one-way communication with third parties may also serve as a disincentive for these parties to file observations. Dominique Guellec and Bruno van Pottelsberghe de la Potterie (2007) claim that after observations are received, the Examination Division adds them to the file and decides whether any of them (which provide better arguments for the case) should be considered, but third parties are not informed of any further action the Division takes in response to their observations. To improve the quality of patents, the EPO should make the search and examination process a forum for lively re-engagement by creating a web-based tool or a funnel for participation through advisory bodies and learned societies that specialise in particular emerging scientific and technological domains. Such mechanisms, would facilitate the discussion, selection and submission of third parties' observations, and provide better opportunities for external actors to provide feedback to the examiners during the pre-grant phase.

In conclusion, it is important to note that an effective search and examination process can contribute to higher quality patents. The EPO should stick to its objectives of promoting innovation by granting valid patents, at a reasonable cost and in due time. To this aim, patent quality mechanisms remain crucial since they require from applicants and examiners to respond effectively to the patent procedures, to contribute to the well-functioning of the patent system and to advancements in innovation. In contrast, the malfunctioning of patent quality mechanisms will serve as an incentive for applicants to rely on socially harmful, strategic behaviour and file broad applications that lack inventive steps and novelty, which would impact patent examiners' workload, lead to huge backlogs and reduce the chances for new emerging technologies to contribute to the innovation process and scientific research.

Notes

1. High-quality patents meet patentability requirements, contribute to the state of the art, offer scientific/social benefit, and stand up to the most rigorous challenges in court.
2. Patent claims are the part of a patent that define the scope of the protection that the inventor seeks in a patent application.
3. Following these examinations, the EPC might provide the applicant with an opportunity to amend those claims that do not fulfill patentability requirements, whereas the Examination Division may conduct an additional examination of these amended claims.
4. For more information on the applicants' filling strategies see also: M. Philipp, 'Patent filing and searching: is deflation in quality the inevitable consequence of hyperinflation in quantity?' *World Patent Information*, vol. 20 (2006), pp. 117–21.
5. Divisional applications are used mostly in cases when the patent application lacks a unity of invention because it describes more than one invention. Thus, the applicant is required to split the patent into several divisional applications each of which claims one single invention.
6. The number of hours spent examining each patent claim fell from 23.8 hours in 1992 to 11.8 hours in 2001.
7. The Administrative Council of the EPO represents all EPC member states. The Board of the Administrative Council was appointed by the Council and set up under Article 28 EPC.

8. 'PACE' is an EPO programme that was established to offer accelerated prosecution of European patent applications, to speed up the search reports and/or patent granting process.
9. 98 comments were submitted by the applicants at the DPMA expressing their thoughts on the UUP.
10. Only 650 third-party observations are filed per year.
11. ISO 9001 is designed to foster quality management and quality assurance.
12. The assessment of the quality of the search and examination process has been conducted based on the internal feedback of the office and the users of the system.
13. 'In collaboration with the European Patent Institution, the EPO has agreed to produce the Manual of Best Practice, which is intended to document the best practices that applicants, their representatives and the EPO should all adopt during the prosecution of an application' Retrieved September 2011 from http://www.epo.org.
14. See also: 'SUEPO Interviews'. Last modified 24 January 2011. Available at http://www.suepo.org/public/interviews.
15. See also section 10.2.
16. Such a system is already established at the USPTO.

References

Akers, N., 'The referencing of prior art documents in European patents and applications', *World Patent Information*, vol. 22 (2000) pp. 309–15.

Alge, D., Future workload 'Study of the Board 28 of the EPO', 2008. Retrieved 10 November 2010 from http://www.ficpi.org/library/08SydneyCET/CET-1404.pdf.

Administrative Council of the EPO, 'Decision of the Administrative Council: amending the implementing regulations to the European Patent Convention and the rules relating to fees', *Official Journal EPO*, (2009), p. 299. Retrieved January 2011 from http://www.epo.org/patents/law/legal-texts/journal/decisions.html.

Borrás, S., C. Koutalakis and F. Wendler, "European agencies and input legitimacy: EFSA, EMeA and EPO in the post-delegation phase", *Journal of European Integration*, vol. 29, (2007), pp. 583–600.

Burke, P. and M. Reitzig, 'Measuring patent assessment quality – analyzing the degree and kind of (in) consistency in patent offices' decision making', *Research Policy*, vol. 36 (2007), pp. 1404–30.

Bessen, J. and M. Meurer, *Patent Failure: How Judges, Bureaucrats and Lawyers Put Innovators at Risk*, New Jersey and Oxford: Princeton University Press (2008).

Brownsword, R., *Rights, Regulation and the Technological Revolution* (New York: Oxford University Press, 2008).

Council of the European Union and Council meeting: Competitiveness, Internal Market, Industry and Research, 2009. Retrieved September 2011 from http://www.consilium.europa.eu.

Cockburn, M., S. Kortum and S. Stern, 'Are all patent examiners equal? Examiners, patent characteristics, and litigation outcomes', in M. Cohen and A. Merill (eds), *Patents in the Knowledge Based Economy* (Washington DC: National Research Council: The National Academic Press, 2003), pp. 1–16.

Cowan, R., W. Eijk, F. Lissoni, P. Lotz, G. Overwalle, and J. Schovsbo, 'Policy options for the improvement of the European patent system', *STOA Report/ FWC/2005-28* (Brussels: European Parliament, 2007).

Edfjäll, C., 'European patent information 2007: EPO policy reformulated', *World Patent Information*, vol. 30 (2007), pp. 206–11.

Elsmore, J. M., 'Quality and quantity: can we have both within the European patent system', *ERA-Forum* 10, 2009, pp. 215–30.

EQS Working Party, *Standard for the European Quality Management System*, 2007. Retrieved January 2011 from http://www.oepm.es/cs.

European Commission, Industrial property rights strategy. COM (2008) 465, DG MARKT, Brussels (2008).

European Patent Office, *EPO Scenarios for the future*, Munich: Druckerei Kriechbaumer (2007).

European Patent Office, Realigning the European patent grant procedure, 2010. Retrieved October 2010 from http://www.epo.org/topics/patentsystem/realigning.html.

Friebel, G., A. Koch, A. Prady and P. Seabright, 'Objectives and Incentives at the European Patent Office', *Institut d'Economie Industrielle Report*, 2006.

Graham, S., B. Hall, D. Harhoff and D. Mowery, *Exploring the Effects of Patent Oppositions: A Comparative Study of US and European Patents.* Presented at the Innovation Policy and the Economy Workshop NBER Summer Institute, 2001.

Guellec, D. and B. van Pottelsberghe de la Potterie, *The Economics of the European Patent System* (New York: Oxford University Press Inc., 2007).

Ganguli, P., 'Intellectual property rights: mothering innovations to markets', *World Patent Information*, vol. 22 (2001), pp. 43–52.

Harhoff, D., *Economic cost-benefit analysis of a unified and integrated European Patent Litigation System*, Germany: EU, Tender No. MARKT/2008/06/D, 2009.

Hammer, Th., 'Bringing the European patent grant process into focus', *Intellectual Asset Management Magazine* (2010), pp. 5–7.

Hunt, R., 'Patentability, industry structure, and innovation', *Journal of Industrial Economics*, vol. 52 (2004) no. 3, pp. 401–25.

Hagel, F., 'Serving two masters: the balance between the applicant and the public', *Patent World*, vol. 16 (2004), pp. 22–4.

Hall, B., 'Patents and patent policy', *Oxford Review of Economic Policy*, vol. 23 (2007) no. 4, pp. 568–87.

Holzer, W., *Patent litigation in Europe – an Adventure.* Seminar on IPR Protection in Europe: Reaching the European Market, July 19–20, 2005, Bangkok.

Jaffe, A and J. Lerner, *Innovation and its Discontents: How Our Broken Patent System is Endangering Innovation and Progress, and What to Do About it* (New Jersey and Oxford: Princeton University Press, 2004).

Kaltenbach, S., S. Krüger, D. Otten-Dünneweber and U. Paschek, *Annual Report: Competence and Quality for over 130 Years*, German Patent and Trademark Office, 2008. Retrieved September 2011 from http://presse.dpma.de/docs/pdf/jahresberichte/korr_jb2003_ engl_def.pdf.

Lee, N., 'Patent eligible subject matter reconfiguration and the emergence of proprietarian norms – the patent eligibility of business methods', *IDEA – The Journal of Law and Technology*, vol. 45 (2005) no. 3, pp. 321–59.

Lemley, M. and C. Shapiro, 'Probabilistic patents', *Journal of Economic Perspectives* (2005), pp. 75–98.

Merrill, A., R. Levin and M. Myers, *A Patent System for the 21st Century* (Washington DC: The National Academies Press, 2004).

Noveck, S., 'Peer to Patent: collective intelligence, open review, and patent reform', *Harvard Journal of Law & Technology*, vol. 20 (2006), pp. 123–61.

Organisation for Economic Cooperation and Development (OECD), *Patents and Innovation: Trends and Policy Challenges* (Paris: OECD, 2004).

Philpott, C., 'A presentation of quality management at the EPO', Power Point Presentation, EPO Directorate General, Lisbon, October 2006.

Scellato, G., F. Caviggioli, Ch. Franzoni, E. Kica, V. Rodriguez, and E. Ughetto, 'Study on the quality of the patent system in Europe', *Official Journal of the European Union*, forthcoming 2011.

Stevnsborg N. and B. van Pottelsberghe de la Potterie, 'Patenting procedures and filing strategies at the EPO', in D. Guellec and B. van Pottelsberghe de la Potterie (eds), *The Economics of the European Patent System: IP Policy for Innovation and Competition* (Oxford: Oxford University Press, 2007).

Schade, J., 'Synergies created by international cooperation in the patent area?', *Competition and Tax Law*, vol. 6 (2009) no. 8, pp. 619–32.

Shang, R., 'Inter partes reexamination and improving patent quality', *Norwestern Journal of Technology and Intellectual Property*, vol. 7 (2009) no. 2, pp. 185–203.

Tang, P., A. John and P. Daniel, 'Patent protection for computer software programs', Report for Directorate General Enterprise, European Commission, 2001. Retrieved 17 November 2010 from http://www.ictlex.net/wp-content/sofstudy.pdf.

Tödtling, F. and M. Trippl, 'One size fits all? Towards a differentiated regional innovation policy approach', *Research Policy, Elsevier*, vol. 34 (2005) no. 8, pp. 1203–19.

Van Pottelsberghe de la Potterie. B., 'Lost property: the European patent system and why it doesn't work', *Bruegel Blueprint Series*, vol. 9 (2009), pp. 3–71.

Wagner, P. R., 'Understanding patent quality mechanisms', *University of Pennsylvania Law Review*, vol. 157 (2009), pp. 1–38.

White, E. K., 'An efficient way to improve patent quality for plant varieties', *Northwestern Journal of Technology and Intellectual Property*, vol. 3 (2004), pp. 9–91.

World Intellectual Property Organization (WIPO), 'World Patent Report: A Statistical Review', 2008. Retrieved 10 December 2010 from http://www.wipo.int/ipstats/en/statistics/pa-tents/wipo_pub_931.html#a11.

11

Regulatory and Policy Impact on the Use of Existing Patents to Enhance Technological Innovation in the Single Market of the European Union

Victor Rodriguez

11.1 Introduction

Patents can be used by industry, government and academia for myriad purposes. What use do patent holders make of their patents? Why are some patents exploited commercially, while others are licensed out to other firms and still others are left unused? These questions are relevant, as the ability to translate new technologies into economically valuable goods or services is crucial for the competitiveness of firms in the single market of the EU.

The aim of this chapter is to review the regulatory and policy impact on the use of existing patents to enhance technological innovation in the EU. The issue of patent use has been studied by scholars (e.g. Guellec, van Pottelsberghe and van Zeebruck, 2007) and patent authorities (e.g. European Patent Office, 2007). Regarding the holder, patents can be used: to restrain the power of suppliers by owning key technology elements; to freeze a technology by preventing the development of a particular market or technology; to set up picket fences through reactive patent behaviour; to prevent others from acquiring IP rights; to create a smokescreen by filing patent applications on technologies which will not be exploited etc. (Guellec, van Pottelsberghe and van Zeebruck, 2007).

In addition, businesses have a variety of reasons for seeking patent protection, which can include: provisional protection of an innovation by holding pending applications; building a monopoly position; blocking others from entering a market; assembling a portfolio of rights to create financial strength; getting a seat at the table when standards are being set; creating marketing messages and becoming more visible in a market; generating licence income; building a base for infringement claims; preventing lawsuits; measuring the performance of the company; communicating innovativeness to investors; avoiding the consequences of not patenting etc. These reasons can be grouped into the following main motivations for patenting: commercial exploitation, licensing, cross-licensing, prevention from imitation, blocking competitors, and reputation. While the uses of patents can be grouped as follows: internal, licensing, cross-licensing, licensing and using, blocking competitors and, not using (European Patent Office, 2007).

The remainder of this chapter is organised as follows. Section 11.2 shows what is already known: What is the knowledge we are building upon? What has already been established? What is the current 'state of the art' in the topic? Section 11.3 presents what is new: What are the recent developments? What is currently being debated? What have we learned from the references? What are the points of contention and what are their implications? Section 11.4 reviews the regulatory and policy impact of the use of existing patents to enhance technological innovation in the EU: What are the future avenues of debate? What are the 'blind spots' that still need to be tackled? Where is the topic or issue headed? Is there a need either for regulatory or policy options or fields that require actions? Are the major issues studied here being left out of the mainstream legislative process? Section 11.5 concludes the chapter.

11.2   What is already known?

This section depicts what is already known: What is the knowledge we are building upon? What has already been established? What is the current 'state of the art' in the topic?

It is unclear how firms are using their patent portfolios and whether an increase in portfolio size leads to an increase in firm innovation, according to Phoebe Chan (2008). For example, Mark Lemley and Carl Shapiro (2005) describe several uses of patents that do not directly lead to the development of a new product. Examples include obtaining financing, boosting market valuation, creating patent thickets, deterring others from suing etc.

As far as statistical evidence is concerned, the European Commission (2004) shows that at the overall EU-6 level: 50.0 per cent of patents are used internally; 35.0 per cent are not used – specifically, 18.7 per cent are filed for strategic reasons and 17.4 per cent are 'sleeping' patents; 15 per cent of patents are exchanged in the market for technologies,

6.4 per cent are licensed, 4.0 per cent are both licensed and used internally and 3.0 per cent are used in cross-licensing agreements. It is worth mentioning that these figures vary across countries, technologies, and applicant size. For instance, the share of unused patents is 18.0 per cent in small and medium-sized enterprises (SMEs) compared to 40.0 per cent in large firms and universities.

With respect to competitors, patents can be used for inventing around or obtaining licences. Apart from that, patents can be used by 'trolls' for litigating patent infringements. Regarding researchers, patents can be used to map the state of the art, to make technological forecasting etc.

Concerning the effectiveness of IP rights protection mechanisms in the formation of research partnerships, patents are most frequently used to protect both background and foreground knowledge in partnerships. Existing patents are quite useful when negotiating new partnerships (Hertzfeld, Link and Vonortas, 2006). In this respect, Bronwyn Hall and Rosemarie Ziedonis (2001) underlined the use of patents as bargaining chips and as a means of avoiding hold-up problems.

Thus large companies and SMEs have different attitudes regarding the use of patents. Patent accumulation concentrated in a few large firms may result in an increase in innovation over time because large firms may be better able to take advantage of economies of scale. Since R&D typically involves large fixed costs, which may include the creation of new patents and the use of existing patents, larger firms may be better equipped to appropriate knowledge through patent ownership. Larger firms may have better access to financing and may be better able to diversify their projects, more effectively decreasing their exposure to risk. Further, significant complementarities often exist among products sold by the same firm. Thus, larger firms may have better knowledge of the demand and size of potential markets (Symeonidis, 1996).

However, if patent ownership is concentrated in the hands of a very few firms, rival firms must subsequently license technology from patent owners or invent around the original patented invention. Therefore, sequential innovation costs for firms in general may rise if future inventions rely upon the previous work of others.

In answer to the questions of how SMEs make use of patents, how SMEs expect to use them and what problems SMEs encounter using the existing patent system, Eurochambres (2006) has said that the main problems relate to high costs, tedious, laborious and time-consuming procedures, language problems, ineffective and costly real protection of patent rights and non-uniformity of systems. As a result, SMEs do not make extensive use of patents. The situation is, however, expected to improve if an effective and cost-efficient patent system is put in place in Europe. Currently, there are two policy steps designed to change the current system: a unitary title and a centralised patent court.

In answering the question – to what extent and for what purpose do innovative firms use patents? – surveys show that patents should be considered as one component in a firm's appropriation strategy, and often not the most important one (Guellec, 2007). To the question of whether or not patents add value to innovation, Ashish Arora, Marco Ceccagnoli and Wesley Cohen (2003) find that for most innovations the patent premium should be negative, which is the reason why so many innovations are not patented and for those which are, the patent premium is significant and has a skewed distribution. To the question of whether or not patents induce further R&D and innovation, the previous authors conclude that patents have a positive impact on R&D expenditure in certain industries.

In particular, patents are used to secure a return from inventions in certain industries. The share of product innovations that are patented is very high for pharmaceuticals, chemicals, machinery, office and computer equipment, and precision instruments but very low for transport and telecommunication services, transport equipment, basic metals, and textile and clothing. Further, patents are more effective for product innovations than for process innovations because processes are not as easily accessible to competitors as products (Arundel and Kabla, 1998). In addition, patents are more often used to protect radical innovations, based on R&D, than to protect more marginal inventions based on other means (Licht and Zolz, 1998).

In addition, the use of patents by research centres and universities is very limited, resulting in a lack of incentives and financing for research, as well as a poor record for licensing and transferring technology and knowledge to industry. The latest European innovation scoreboard listed only five European countries as innovation leaders. This shows that patenting issues remain insufficiently addressed (Pompidou, 2007).

All in all, this section has shown the various uses of patents by various actors in the patent system and the extent of such uses.

11.3 What is new?

This section presents what is new: What are the recent developments? What is currently being debated? What have we learned from the references? What are the points of contention and what are their implications?

Increasing competition provides the incentive for firms to become more mobile, moving to regions where labour costs are low, the regulatory burden is less onerous and conditions are negotiable. Consequently, companies are increasingly turning to international expansion and foreign direct investment (European Patent Office, 2007).

Foreign direct investment (FDI) is sensitive to international differences in IP rights in sectors with knowledge-based assets. FDI representing complex but easily copied technologies is likely to increase with stronger IP rights, because patents increase the value of knowledge-based assets, which may be efficiently exploited through internalised organisation. To the extent that licensing costs come down with stronger IP rights, FDI could be displaced over time by efficient licensing. Whatever the mode, the likelihood that the most advanced technologies will be transferred rises with strengthened IP rights. Investment and technology transfers are relatively insensitive to international differences in IP rights in sectors with old products and standardised, labour-intensive technologies. Here, FDI is influenced by factor costs, market sizes, trade costs and other location advantages (Maskus, 2000).

Royalties and licence fees are affected by the strength of patent regimes. IP rights can underpin an efficient system of contracts to promote formal technology transfers through licensing. The potential increases in licensing volumes from strengthening such rights could be significant, and the quality of the technologies should rise. In this respect, technology importers will pay higher costs to absorb more and better technologies as a result of tighter IP rights. Stronger IP rights could also permit firms to choose not to license their closely held technologies except in cross-licensing or patent-pooling arrangements. Thus, a trade-off is likely to emerge between stronger licensing incentives and greater prerogatives to maintain technologies under close control (Yang and Maskus, 2001).

The use of existing patents would enhance technological innovation in the EU, if there were a centralised jurisdiction and a unitary title in the EU single market. By doing this, the system would provide more incentives to use European patents and to foster innovation and competitiveness in the knowledge economy in Europe.

Contrasting the evidence with recent initiatives, the European and EU Patent Court have drafted aims to create a centralised European patent litigation system to avoid a multiplicity of parallel national litigations and thus to drive down litigation costs. In addition, the European and EU Patent Court draft aims to offer a solution to the current problem of contradictory outcomes and thus to increase legal certainty in the European patent system.

The centralised patent court proposal was itself a compromise, between those who favoured a specialised patent court outside of the EU legal structure and those who said that it must come squarely under the jurisdiction of the European Court of Justice (ECJ) (Nurton, 2011c).

Institutionally, the Council of the European Union (2009) agreed on the features of the European and EU Patent Court. The Council stressed that the conclusions are without prejudice to the request for an opinion from the ECJ as well as to member states' individual written submissions, and are subject to the opinion of the ECJ.

In March 2011, the ECJ said that the proposed agreement on the European and EU Patent Court would deprive EU member states and the ECJ of their powers to interpret EU law. In its opinion, the ECJ first set out that the request was admissible. It then addressed the substance of the arguments. The ECJ said that Articles 262 and 344 of the EU Treaty (TFEU or Lisbon Treaty) cannot preclude the creation of the European and EU Patent Court. But it then stressed that the essential characteristics of EU legal order are 'its primacy over the laws of the Member States' and 'it is for the ECJ to ensure respect for the autonomy of the EU legal order thus created by the Treaties'. It is for national courts and tribunals and the ECJ itself to ensure the full application of EU law in all EU member states 'and to ensure judicial protection of an individual's rights under that law'. But the proposed European and EU Patent Court 'is outside the institutional and judicial framework of the EU' and has a distinct legal personality under international law. '[T]he Member States cannot confer the jurisdiction to resolve [...] disputes on a court created by an international agreement which would deprive [...] 'ordinary' courts [...] to implement EU law and, thereby, of the power provided for in Article 267 TFEU, or, as the case may be, the obligation, to refer questions for a preliminary ruling in the field concerned', said the opinion. 'It is clear that if a decision of the [European and EU Patent Court] were to be in breach of EU law, that decision could not be the subject of infringement proceedings nor could it give rise to any financial liability on the part of one or more Member States', added the opinion (Nurton, 2011a).

The Administrative Council of the European Patent Organisation has agreed to vastly increase investment in the current cooperation programmes between the EPO and member states' NPOs in order to accelerate the optimisation of automated translation systems.[1] As far as quality is concerned, the development of 'fit-for-purpose' machine-translation technology not only enables a technically qualified user skilled in the art to understand the technical content of the patent document (fit-for-purpose) but also enables a technically qualified user skilled in the art to assess whether a given patent document is relevant from a technical or economic point of view (minimum quality).[2]

According to the Council of the European Union's (2009) conclusions on an enhanced patent system for Europe, the EU Patent Regulation should be accompanied by a separate regulation, which should govern the translation arrangements for the EU patent adopted by the Council with unanimity in accordance with the TFEU (or Lisbon Treaty). The EU Patent Regulation should come into force together with the separate regulation on translation arrangements for the EU patent.

In particular, the renewal fees for EU patents should be progressive throughout the life of the patent and, together with the fees paid during the application phase, cover all costs associated with the granting and administration of the EU patent. The renewal fees would be payable to the EPO, which would retain 50 per cent and distribute the remaining amount among the member states in accordance with a distribution key to be used for patent-related purposes.

Furthermore, a Select Committee of the Administrative Council of the European Patent Organisation should fix both the exact level of the renewal fees and the distribution key for their allocation once the EU Patent Regulation enters into force. The Select Committee should be composed of representatives of the EU and of all the member states. The position to be taken by the EU and the member states in the Select Committee would need to be determined within the Council, at the same time as the EU Patent Regulation is adopted. The level of renewal fees should, in addition to the above-mentioned principles, be fixed with the aim of facilitating innovation and fostering the competitiveness of European business. It should also reflect the size of the market covered by the EU patent and be similar to the level of the renewal fees for what is deemed to be an average European patent at the time of the first decision of the Select Committee.

In addition, the distribution key should take into account a selection of fair, equitable and relevant criteria such as, for instance, the level of patent activity and the size of the market. The distribution key should provide compensation for, among other things, having an official language other than one of the official languages of the EPO, for having disproportionately low levels of patent activity and for more recent European Patent Convention, EPC, membership.

While some delegations from EU member states would prefer to keep progress on the centralised patent court separate from that on the unitary title, others are of the opinion that consensus should be reached on both areas simultaneously. As regards the EU patent, discussions have focused on the two main outstanding issues, viz. translation arrangements and the distribution of revenue from renewal fees. It is felt that an agreement on these two issues would considerably facilitate an overall agreement on the EU Patent Regulation. There was broad agreement that in the interest of the users of the patent system, in particular SMEs, EU patents must be affordable. With respect to translation arrangements, a majority of delegations from the member states would welcome the idea of exploring solutions making use of automated translation systems. Regarding the distribution to NPOs of part of the revenue accruing from renewal fees for EU patents, appropriate criteria were suggested, taking into account the size of the market, the level of patent activity and improving the access of SMEs to the patent system (Council of the European Union, 2008).

Negotiations were needed on the EU patent and special arrangements were crucial for the language. Officials from EU member states have discussed the potential of using machine translations of patent documents for the proposed EU patent. Although the Lisbon Treaty cut the number of decisions that must be reached on the basis of unanimity, this issue is one that all EU member states must agree on. The language issue proved to be a big hurdle in achieving a final deal since Spain and Italy did not agree that patents needed only be translated into English, French and German.

The Council of the European Union (2011) authorised enhanced cooperation on the creation of unitary patent protection, 25 of the EU member states backed 'enhanced cooperation' for the EU Patent. Enhanced cooperation is a procedure allowing a number of EU states to proceed with legislative plans in cases where unanimity is not possible. When introduced, the unitary patent would apply on grant in all 25 countries that have opted-in, without further validation. It would be administered and examined by the EPO, which will maintain the EU patent registry and process renewal fees (Nurton, 2011b).

All in all, this section has presented the latest institutional developments to enhance technological innovation in the EU single market.

11.4   Regulatory and policy impact

This section reviews the regulatory and policy impact on the use of existing patents to enhance technological innovation in the EU: What are the future avenues of debate? What are the 'blind spots' that still need to be tackled? Where is the topic or issue headed? Is there a need either for regulatory or policy options or fields that require actions? Are the major issues studied here being left out of the mainstream legislative process?

Previous sections have shown that the use of patents is associated with the regime. It was added that the use of existing patents would enhance technological innovation in the EU, if a centralised jurisdiction and a unitary title were to exist in the EU single market.

If a unitary title regime were to exist in the European patent system, then the granting authority would be the EPO. If a centralised jurisdiction were to exist, then that body would the European and EU Patent Court. These two regime constructions can be debated at various levels of analysis. At the macro-level, such unitary title and centralised jurisdiction can be discussed using the inter-governmentalist theory or realism's sovereignty. First, member states delegate authority to the EPO to enable it to grant the unitary title. However, delegation of the tasks of grating unitary titles implies no loss of autonomy. Second, member states delegate authority to the centralised court to enable it to monitor the enforcement of patent rights. However, this delegation to the centralised court implies no loss of autonomy since the principles governing the centralised court's decisions were unanimously approved by member states during the Swedish Presidency in 2009.

Further, such a unitary title and centralised jurisdiction construction can also be discussed by the principle of fiscal federalism. First, patent granting is assigned to the supra-national body, some other patent-related tasks are shared between the supra-national and national level, and other tasks are set out exclusively at national level. Second, Patent enforcement is assigned to the EU level, some other enforcement tasks are shared between the EU and member states, and other tasks are set out exclusively by member states. When assigning tasks to the supra-national body, the EU embraced the subsidiarity, proportionality and added value principles. Subsidiarity means that the EU should not take actions unless doing so is more effective than action taken at national, regional or local levels. Proportionality means that when the objectives can be better achieved at the EU level, the EU should undertake only the minimum necessary actions. European added value means that European policies require a synergy effect from such supra-nationality not attainable within national borders.

At the meso-level, the development of such unitary title and centralised jurisdiction construction reflects neo-liberal institutionalist theory, or functionalism, which predicts that national governments develop international institutions to capture perceived gains where the benefits exceed the costs. First, validation and translations at national level represent a shortcoming of the fragmented European patent system. If a unitary title were to respect Community legal order, co-existence with European and National Patents, affordability, cost efficiency, legal certainty, high quality, non-discrimination, a pre-grant phase regulated by the EPC, and a post-grant phase regulated by the Community Patent Regulation, then it would solve the problem of costs and uncertainty. Second, enforcement at the national level represents a shortcoming of the fragmented European patent system. If a unified jurisdiction were to centralise the patent enforcement regime, then it would solve the problem of costs and uncertainty.

At the micro-level, such a unitary title and centralised jurisdiction construction can also be discussed by the principal-agent theory. First, member states are regarded as the principals and the granting patent office is considered as the agent. If the principals consider delegation for mere efficiency, then the patent authority is considered just an executor, which is expected to diminish the costs of multiple validations and translations. Second, member states are regarded as the principals and the centralised court is considered to be the agent. There exist two different roles to the centralised enforcement and the member states. If the principals consider delegation for mere efficiency, then the centralised court is considered to be just an executor, which is expected to diminish the costs of multiple litigations. If the principals consider delegation in order to strengthen enforcement, then the centralised court is playing the role of a custodian, whose task is to eliminate any diverging court outcomes.

All in all, this section has reviewed these two regime constructions at macro, meso and micro-level of analysis.

11.5   Conclusion

The use of patents is associated with the regime of technological innovation. The use of existing patents would enhance technological innovation in the EU, if a centralised jurisdiction and a unitary title were to exist in the EU single market.

The current European patent regime allows the retention of institutional arrangements within member states and prevents any moves to delegate responsibility outside the national sphere. Such a regime is characterised by a fragmented European patent system of national translation, validation and enforcement. Fragmentation is regarded as a failing of the system due to higher costs and uncertainty, and low quality patents. How can problems caused by such fragmentation be solved institutionally?

There is no single approach, but at least two options can be constructed:

(A) Either make a unitary title and a centralised patent court legitimate to enhance the competitiveness of the European single market and innovation in the knowledge economy.

This would provide policy coherence and cohesion by defining what the regime is and what it does in relation to the economy. Agreements would succeed with a conviction of why Europe is lagging behind in terms of innovation and competitiveness in the knowledge economy. Over domestic negotiations, EU member states would absorb the concern of domestic actors and build coalitions with them. At the European level, EU member states would try to implement these concerns without committing to anything that would have deleterious effects at home. A 'quid pro quo' approach with concessions in areas outside patent regime might offer some room for manoeuvre in making progress in the negotiations towards an agreement.

(B) Or allow for parallel regimes when there is a tension between strongly institutionalised differences across EU member states, and the desire of policy makers and stakeholders to adapt common rules for mutual advantage in the European patent system.

In addition, there are three other options possible for patent enforcement. The first is to integrate the centralised patent court into the ECJ system as a new chamber of patent specialists who will decide without delays, uncertainty and confusion. However, this would-be jurisdiction would need to be accepted by the non-EU countries of the EPC. The second option would be to resume negotiations over the European Patent Litigation Agreement, EPLA, for a mutual recognition of European patent judgments, excluding EU patents, provided that EU law issues are referred to the ECJ. However, this would-be jurisdictional arrangement would need to be accepted by some EU member states, led by France. The third option would be to keep the 'status quo' with national courts, where 90 per cent of litigation takes place in four jurisdictions (i.e. Germany, United Kingdom, France and the Netherlands).

One way of giving a chance to EU member states that support a unitary title and centralised patent enforcement is by means of enhanced cooperation, which allows those member states that wish to continue to work more closely together to do so. The advantage of enhanced cooperation is that it does not require unanimity from EU member states. Parallel regimes and enhanced cooperation are therefore a strategy but not a final outcome. The ultimate goal would be to create a single institutional architecture by means of a unitary title and a centralised patent jurisdiction for the single market of the EU provided that patent-related costs were diminished and certainties were raised. It is suggested that more EU member states progressively abandon their national systems to this end.

Notes

1. Available at http://documents.epo.org/projects/babylon/eponet.nsf/0/ E8E346B6B89D903DC12577D600311DBE/$File/battistelli_speech_20101108. pdf. Accessed on 21 September 2011.
2. Available at http://documents.epo.org/projects/babylon/eponet.nsf/0/ FDB7AEB9AB8D5336C12577CA0050972D/$FILE/GeorgArtelsmair_MT_ developments.pdf. Accessed on 21 September 2011.

References

Arora, A., M. Ceccagnoli and W. Cohen, 'R&D and the patent premium', *NBER Working Paper 9431*, 2003.

Arundel, A. and I. Kabla, 'What of percentage of innovations are patented? Empirical estimates for European Firms', *Research Policy*, vol. 27 (1998), pp. 127–41.

Chan, P., 'Does sequential innovation decline in the presence of large patent portfolios? Examining future variety creation in the U.S', *Agricultural Biotechnology Industry*, (2008). Available at http://papers.ssrn.com/sol3/papers. cfm?abstract_id=1008970. Accessed on 21 September 2011.

Council of the European Union, 'Enhancing the patent system in Europe', 2008. Available at http://register.consilium.europa.eu/pdf/en/08/st09/st09473.en08. pdf. Accessed on 21 September 2011.

Council of the European Union, 'Press Release of the 2982 Council Meeting. Competitiveness (Internal Market, Industry and Research)', 2009. Available at http://www.consilium.europa.eu/uedocs/cms_data/docs/pressdata/en/ intm/111732.pdf. Accessed on 21 September 2011.

Council of the European Union, 'Council authorises enhanced cooperation on creation of unitary patent protection', 2011. Available at http://www. consilium.europa.eu/uedocs/cms_data/docs/pressdata/en/intm/119732.pdf. Accessed on 21 September 2011.

Eurochambres, 'Position on future patent policy in Europe', 2006. Available at http://ec.europa.eu/internal_market/indprop/docs/patent/hearing/tilman_ en.pdf. Accessed on 21 September 2011.

European Commission, 'Study on evaluating the knowledge economy: what are patents actually worth? The value of patents for today's economy and society', 2004. Available at http://ec.europa.eu/internal_market/indprop/docs/patent/ studies/patentstudy-summary_en. Accessed on 21 September 2011.

European Patent Office, 'Scenarios for the future', 2007. Available at http:// documents.epo.org/projects/babylon/eponet.nsf/0/63A726D28B589B5BC1

2572DB00597683/$File/EPO_scenarios_bookmarked.pdf.   Accessed   on 21 September 2011.

Guellec D., B. van Pottelsberghe and N. van Zeebruck, 'Patent as a market instrument', in D. Guellec and B. van Pottelsberghe (eds), *The Economics of the European Patent System: IP Policy for Innovation and Competition* (Oxford: Oxford University Press, 2007).

Guellec, D., 'Patent as an incentive to innovate', in D. Guellec and B. van Pottelsberghe (eds), *The Economics of the European Patent System: IP Policy for Innovation and Competition* (Oxford: Oxford University Press, 2007).

Hall, B. and R. Ziedonis, 'The patent paradox revisited: an empirical study of patenting in the US semiconductor industry', *Rand Journal of Economics*, vol. 32 (2001), pp. 101–28.

Hertzfeld, H., A. Link and N. Vonortas, 'Intellectual property protection mechanisms in research partnerships', *Research Policy*, vol. 35 (2006), pp. 825–38.

Lemley, M. and C. Shapiro, 'Probabilistic patents', *Journal of Economic Perspectives*, vol. 19 (2005), pp. 75–98.

Licht, G. and K. Zolz, 'Patents and R&D: an econometric investigation using applications for German, European and US patents by German companies', *Annales d'économie et de statistique*, vol. 49 (1998), pp. 329–60.

Maskus, K., *Intellectual Property Rights in the Global Economy* (Washington: Institute for International Economics, 2000).

Nurton, J., 'Court deals blow to EU patent court', *Managing Patents*, 2011a.

Nurton, J., 'Green light for enhanced cooperation', *Managing Patents*, 2011b.

Nurton, J., 'Analysis: EU patent – nice bullets, no gun', *Managing Patents*, 2011c.

Pompidou, A., 'The role of the European Patent Office in European patent policy: a Europe of innovation – fit for the future?', 2007. Available at http://www.docstoc.com/docs/40453511/President-European-Patent-Office. Accessed on 21 September 2011.

Symeonidis, G., 'Innovation, firm size and market structure: Schumpeterian hypotheses and some new themes', *OECD Economics Department Working Papers*, Number 161, 1996.

Yang, G. and K. Maskus, 'Intellectual property rights and licensing: an econometric investigation', *Review of World Economics*, vol. 137 (2001), pp. 58–79.

Conclusion

Michiel A. Heldeweg and Evisa Kica

The contributors to this book have attempted to develop a multidisciplinary approach towards the role of regulation in technological innovation. Far from providing a uniform solution on how regulation is responding, or should respond, to technological developments, the contributors have presented a spectrum of related arguments and opinions. Clearly, the emergence of new technologies, the complex blends of public and private actors and new trans-governmental networks within the regulatory space have created a major regulatory challenge for the current technology trends in various sectors. In this regard, 'regulation of innovation is by definition dynamic, and certainly does not deserve a one size fits all approach'!

In line with the findings and opinions voiced by the 'smart regulation movement', the innovation and regulation 'relationship' demonstrates a shift from government to governance. Not only does the regulation of innovation need to be responsive to the specific characteristics of innovation opportunities and technological narratives, and thus be sensitive to the possibilities of tailoring regulation as a mixture of concepts and forms, but it also clearly bears witness to the need to connect public and private stakeholders through and towards regulatory interaction. When it comes to innovation, 'regulatory governance' involves both regulation and governance. This observation is particularly important given the fact that the open dynamics of technological innovation require a shared awareness from all stakeholders as to the opportunities and risks involved, and a shared need to accommodate change while retaining sufficient mutual trust and certainty.

Further, an important aim of this book is to support and elucidate regulatory relevance as regards fostering innovation while, at the same time, considering the need for regulation to strike a balance between fostering innovation and protecting against technological risks. In the context of this book, the element of fostering innovation refers strongly to the Lisbon agenda and to the observation that the advancement of innovation through regulation is scientifically underexplored and not well understood.

To rectify this state of affairs, we have argued that a multidisciplinary approach is needed if regulation is effectively to foster and secure technological innovation. In this book, we have endeavoured to present such a multidisciplinary approach through a range of contributions that are grouped in three sections. The first section set out to provide a general analysis and appraisal of the regulation and innovation relationship, featuring such comparative advantages as legal designs and informal regulation. The second section was built around specific sectors (in particular telecommunication infrastructures and related services, competition law and regulated industries, border management, energy innovation and PPP), with a focus on issues related to regulating technological innovation. The final section brought forward the much-debated issue of emerging technologies and the challenges they present to innovation and regulation. In this section, the issues of regulatory partnerships in nanotechnology, patent quality and the use of existing patents were debated.

Besides generating a call for greater awareness of innovation and regulation (an ambition that we believe all the authors have met) the chapters provide various methodological insights and approaches towards analysing the technology revolution and the regulatory challenges in specific sectors, emerging technologies and interests. In searching for the optimal solution to encourage effective technological innovation, the contributors' suggestions, as taken from their methodological insights, are twofold. First, that regulation develops as technology advances and revolutionises, which implies a need to open up science and technology development, foster a balance between state power and public scrutiny, and aim for smart rules and regimes. Second, by focusing on the process of patenting emerging technologies and regulating nanotechnology, the need for an *ex ante* regulatory design is highlighted. New technologies are no longer regarded as entirely passive entities that are developed by inventors and used by customers; rather, they involve complex networks of actors and outcomes, and regulation should act to control and limit the effects of such technology developments.

In the introduction we presented three critical questions that this book addresses:

- what type of regulatory framework would best fit the needs of technology and innovation developments?

- what competences or authorities should be given to the regulatory actors and other stakeholders to shape the future paths of technology innovation?
- what lessons can we distil from other regulatory fields, and how we can apply what we have learnt to further enhance the development of technology innovation?

Although not intended to capture exclusively this book's subject matter, it makes sense to reflect on these questions and on how the authors and the overall book have approached these issues and what is seen as the way forward with regard to the challenges in regulating technological innovation.

As to the first question (What type of regulatory framework would best fit the needs of technology and innovation developments?), we feel that our observation in the above preamble stands: that in regulating technology innovation there is no single innovation-specific type of regulatory framework that can be expected to respond adequately to the full diversity found in technological and innovative sectors. Having said this, across the various contributions to this book it becomes clear that regulators' responsiveness to 'innovation specificity', dealing with related dynamics and conflicting interests, must be matched with legal certainty, both with regard to companies that aim to invest in innovation (while securing a return on their investment) and to consumers and third parties who aim to be free from hazardous technological risks (and desire adequate and enforceable technical standards). Such certainty could be provided in various regulatory forms and procedures, some of which may place government centre stage (e.g. as legislator or leading customer), but others where government merely ensures the basic rules of the game, within which private actors determine innovative pathways. Clearly, the design and introduction of regulation that fosters innovation must be subject to both *ex ante and ex post* impact assessments, as regulation is inevitably a major component of the institutional space that allows innovation to flourish.

Regarding the second question (What competences or authorities should be given to the regulatory actors and other stakeholders to shape the future paths of technology innovation?), it is clear that a 'smart' broadening of the regulatory scope must be recognised. While many of the above remarks still bear witness to government setting the stage, in practice we have clearly moved from government to governance including in terms of regulatory engagement with innovation. The element of informal law, or of self or private regulation, is ubiquitous in innovation narratives. With this observation in mind, the matter of competences and authority to regulate must be seen from a different perspective. Here, not so much the traditional *Rule of Law* but rather legal pluralism and regulatory constitutionalism become the new challenges in regulatory governance. This is a profound challenge, not merely in finding appropriate alternatives to replace previous safeguards of legitimacy and effectiveness, but also because innovation cannot thrive if the institutional surroundings are unstable and unpredictable in terms of the state of play they project (such as a level playing field), their enforcement and possibilities for change. Moreover, innovation is at risk if there are no genuine constraints on the possibility of an ever-expanding hybrid regulatory web, with corresponding embedded interests and administrative burdens.

Finally, with regard to our third question (What lessons can we distil from other regulatory fields, and how we can apply what we have learnt to further enhance the development of technology innovation?), we have seen that the regulation of technological innovation builds upon wider regulatory theory – references to which were made in the introduction to this book. Against this backdrop, the various contributors have touched upon a variety of fields – such as telecommunications, health and safety, intellectual property, environment, energy and nanotechnology – within which the innovation narrative is studied from a regulatory angle. Clearly, these analyses show that, although each narrative comes with its own regulatory specificities, comparisons are important as with technological innovation comes social innovation, and social institutions must or will change in the wake of technology thrusts. The revolution in innovation and more specifically the development of new technologies involve high levels of uncertainty (uncertainty over the development of these technologies, over their uncontrolled introduction into process technology and consumer products and over difficulties in providing effective private regulatory supervision) and it is important for regulation to become a key factor in establishing the necessary social innovation, and to safeguard a careful transition to, and practice of, new inventions while considering more than mere technological interests. The chapters in this book make clear that by establishing an inclusive regulatory design of information flow, and an open interchange among stakeholders on the burdens and benefits of technological innovation, the influence and the acceptance of these innovative developments will be increased, benefiting both public and private actors.

As editors, we feel that this book is a step towards enhancing our understanding of the regulation and innovation relationship. We are convinced that such an ambition requires a multidisciplinary approach as innovation narratives come with a high complexity, reflected in the relationships between the technological and social mechanisms involved, which need to be sufficiently understood to employ regulation successfully to advantage innovation. We hope that readers appreciate our efforts and we look forward to their opinions.

Index

Page numbers followed by n indicate notes.